ILLUSTRATED CHANGES IN THE NATIONAL ELECTRICAL CODE®

Third Edition

**Based on the 1999
National Electrical Code®**

Ronald P. O'Riley

Delmar Publishers

an International Thomson Publishing company

Albany • Bonn • Boston • Cincinnati • Detroit • London • Madrid
Melbourne • Mexico City • New York • Pacific Grove • Paris • San Francisco
Singapore • Tokyo • Toronto • Washington

NOTICE TO THE READER

Cover design courtesy of: Brucie Rosch

Delmar Staff
Publisher: Alar Elken
Acquisitions Editor: Mark Huth
Project Editor: Megeen Mulholland
Production Coordinator: Toni Bolognino
Art and Design Coordinator: Cheri Plasse
Editorial Assistant: Dawn Daugherty

COPYRIGHT ©1999
By Delmar Publishers
an International Thomson Publishing Company

The ITP logo is a trademark under license.
Printed in the United States of America

Online Services

Delmar Online
To access a wide variety of Delmar products and services on the World Wide Web, point your browser to:
http://www.delmar.com
or email: info@delmar.com

A service of I(T)P®

For more information, contact:

Delmar Publishers
3 Columbia Circle, Box 15015
Albany, New York 12212-5015

International Thomson Publishing Europe
Berkshire House
168-173 High Holborn
London, WC1V7AA
United Kingdom

Nelson ITP, Australia
102 Dodds Street
South Melbourne,
Victoria, 3205 Australia

Nelson Canada
1120 Birchmont Road
Scarborough, Ontario
M1K 5G4, Canada

International Thomson Publishing France
Tour Maine-Montparnasse
33 Avenue du Maine
75755 Paris Cedex 15, France

International Thompson Editores
Seneca 53
Colonia Polanco
11560 Mexico D. F. Mexico

International Thomson Publishing GmbH
Königswinterer Strasße 418
53227 Bonn
Germany

International Thomson Publishing Asia
221 Albert Street
#15-01 Albert Complex
Singapore 189969

International Thomson Publishing Japan
Hirakawa-cho Kyowa Building, 3F
2-2-1 Hirakawa-cho, Chiyoda-ku
Tokyo 102, Japan

ITE Spain/Paraninfo
Calle Magallanes, 25
28015-Madrid, España

2 3 4 5 6 7 8 9 10 XXX 04 03 02 01 00 99 98

Library of Congress Cataloging-in-Publication Data
O'Riley, Ronald P., 1914–
   Illustrated changes in the National electrical code / Ronald P.
O'Riley. — 3rd ed.
      p.   cm.
   "Based on the 1999 national electrical code."
   Previous eds. published 1993 under title: Illustrated changes in
the 1993 National electrical code; and 1996 under title: Illustrated
changes in the 1996 National electrical code.
   ISBN 0-7668-0763-0 (alk. paper)
   1. National Fire Protection Association. National Electrical Code
(1999) 2. —Electric engineering—Insurance requirements—United
States. I. O'Riley, Ronald P., 1914– . Illustrated changes
in the 1996 National electrical code. II. National Fire Protection
Association. National Electrical Code.
TK260.059 1998
621.319'24'0218—dc21   97-17960
                                                      CIP

CONTENTS

PREFACE

■ ■

The purpose of *Illustrated Changes in the National Electrical Code®* is to help inform the electrical industry and other interested parties of the many revisions, deletion,s and rearrangements made in the 1999 edition of the *National Electrical Code® (NEC®)*. (*National Electrical Code®* and *NEC®* are registered trademarks of the National Fire Protection Association, Inc., Quincy, Massachusetts.)

All analyses and comments are those of the author and are not to be construed as official interpretation. The Authority Having Jurisdiction (AHJ) has the final authority for interpretation as per Section 90-4 of the *Code*. Illustrations, reasons for changes, field applications, and analytical comments are included as aids for understanding the changes. The cross checks presented are the author's. They may not always agree with the *Code*, but are given as a general reference to show where things are moved from and to illustrate some of the new section identification numbers.

This guidebook can be used as a ready reference for individual study or classroom study. A set of transparency masters of the illustrations in the text is also available from Delmar Publishers.

The guidebook chapter, article, and section numbering sequence follow that of the 1999 *Code*. Each change is identified as new, revised, or deleted text. Where an exception is revised or a new exception is added, the applicable basic rule immediately precedes the exception for ready reference.

The illustrations in this text contain icons calling attention to the nature of the changes. For a revision, the icon is a flag with the word REVISED on it. This means the basic core of the rule remains, but some important adjustment has been made. For something new, the icon indicates a NEW ARRIVAL. A new arrival may be something completely new or it could be such a radical revision that there will be a major change if an installation is made. For a deletion, the BOOT icon shows that something has been removed from the *Code*, often making a big change in the method of installation. The 1999 *Code* moves many things. Therefore, a MOVER truck icon calls attention to their relocation. Many of these moves do not change the *Code* rule, but simply place it in a different section. A THUMBS-DOWN icon indicates something that the AHJ will be looking at as a violation. A THUMBS UP icon indicates an acceptable installation

Keeping up to date with the fast-changing electrical industry is a constant challenge for everyone involved, from the design engineer and the journeyman to the apprentice and the electrical inspector. Many new electrical construction materials, new products, and methods of installation are constantly being discovered. The *Code* is revised every three years to keep up with the changing electrical industry. This guidebook is a teaching aid that enables electricians and others to quickly understand and implement the changes.

INTRODUCTION

■ ■

KEY TO THE USER-FRIENDLY
1999 NATIONAL ELECTRICAL CODE®

One outstanding goal in the revising of the 1999 *National Electrical Code®* is to make it more *USER-FRIENDLY.*

The various code-making panels, task groups and ad hoc committees have worked hard and long toward this goal, and their efforts shine in the 1999 *Code*. The 1999 edition of the *Code* is more **user-friendly** for the inspector as to enforceability; more **user-friendly** for the local authorities having jurisdiction (AHJ) who adopt it, and more **user-friendly** for the rank-and-file journeymen in the field.

With this aim in mind, it is easier to accept and understand the multitude of rule consolidations and rearrangements that take place in the 1999 *Code*. The electrical industry kept marching on toward the future, and the 1999 *Code* had to be brought into line with new developments. All the moving about and renumbering of sections may give the seasoned user a little problem at first. But for the newcomer and the average user, it will be easier to use.

Before starting into the changes, it would be well to look at some of the operations used to accomplish this **user-friendly** goal.

Key Thought

Watch how exceptions are reworded and moved into the main body of the section without altering the intent of the rule.

A very large number of exceptions have been restated in a positive way and made a part of the main body of the text, or subdivided under the main section heading and given boldface identification. To accomplish this, the *Code* has introduced a new term in Section 90-5(b) "Permissive Rules". This section is a good example of how rearrangement takes place.

Cross Check

1996	1999

90-5 Mandatory Rules and Explanatory Material
    Section consisted of one sentence
	90-5 Mandatory Rules, Permissive, Rules, and Explanatory Material
	    (a) Mandatory Rules (More detailed)
	    (b) Permissive Rules. (All new)
	    (c) Explanatory Material
	        (More detailed)
	        FPN (New)

Several words or phrases are used in conjunction with **permissive rules**. These include "shall not be required," "shall be permitted," "unless," "except," and "excluding." Examples are as follows.

Making a positive statement out of an exception by the use of the words *"excluding"* and *"unless."*

Comparing

1996	1999

200-10 Identification of Terminals.
 	
   (a) Device Terminals. All devices provided with terminals for the attachment of conductors and intended for connection to more than one side of the circuit shall have terminals properly marked for identification.
Exception No. 1: Where the electrical connection of a terminal intended to be connected to the grounded conductor is clearly evident.
Exception No. 2: The terminals of lighting and appliance branch-circuit panelboards.
Exception No 3: Devices having a normal current rating of over 30 amperes other than polarized attachment plugs and polarized receptacles for attachment plugs as required in Section 200-10(b).
	200-10 Identification of Terminals.
	
	   (a) Device Terminals. All devices, *excluding lighting and appliance branch-circuit panelboards* (was Ex. 2) provided with terminals for the attachment of conductors and intended for connection to more than one side of the circuit shall have terminals properly marked for identification, <u>**unless**</u> **the electrical connection of the terminal intended to be connected to the grounded conductor is clearly evident** *(Was Ex. 1).*
	Exception: Reword with no change in intent

Revision Keys

When an exception is applied, the exception ONLY applies to the rule directly preceding the exception. If the exception applies in a second place, it is repeated. An example 1999 *Code*:

210-8(a)(5) GFCI receptacles required in unfinished basements.
   Exception No. 1 Receptacles that are not readily accessible.

   This exception applies only to Section 210-8(a)(5) and cannot be applied to the preceding Section 210-8(a)(4) for Crawl Spaces or Section 210-8(a)(3) for Outdoors. Exceptions cannot be taken out of context. If the same exception is needed in another location, the exception is repeated.

210-8(a)(2) GFCI Receptacles are required in garages and unfinished accessory buildings.
   Exception No. 1 Receptacles that are not readily accessible.

   Note how the two exceptions to 210-8(a)(2) and 210-8(a)(5) are identical but are applied to two different locations and are therefore repeated.

Key Thought: Understanding the Rearrangement of Rules

The following is an example of how a long, complicated section with a multitude of exceptions is rearranged and blended into the text with boldface headings. Now, the separate rules can be more quickly located and clearly understood.

Crosscheck

1996	1999

250-23 Grounding Service-Supplied Alternating-Current Systems
(a) System Grounding Connections
A premises wiring system that is supplied by an ac service that is grounded shall have at each service a grounding electrode conductor connected to a grounding electrode that complies with Part H of Article 250. The grounding electrode conductor shall be connected to the grounded service conductor

at any accessible point from the load end of the service drop or service lateral to and including the terminal or bus to which the grounded service conductor is connected at the service disconnecting means.

(FPN) Moved up from end of paragraph

Where the transformer supplying the service is located outside the building, at least one additional grounding connections shall be made from the grounded service conductor to a grounding electrode, either at the transformer or elsewhere outside the building.

Exception No. 6: As covered in Section 250-27 for high-impedance grounded neutral systems grounding connection requirements.

Exception No. 4: For services that are dual fed (double ended) in a common enclosure or grouped together in separate enclosures and employing a secondary tie, a single grounding electrode connection to the tie point of the grounded circuit conductor from each power source shall be permitted.

250-24 Grounding Service-Supplied Alternating Current Systems
(a) System Grounding Connection
A premises wiring system that is supplied by an ac service that is grounded shall have at each service a grounding electrode conductor connected to a grounding electrode*(s) as required by* Part C of Article 250. The grounding electrode conductor shall be connected to the grounded service conductor *in accordance with (1) through (5).*

(1) General. The connection shall be made at any accessible point from the load end of the service drop or service lateral to and including the terminal or bus to which the grounded service conductor is connected at the service disconnecting means.

(FPN) Relocated—no change

(2) Outdoor Transformer. Where the transformer supplying the service is located outside the building, at least one additional grounding connections shall be made from the grounded service conductor to a grounding electrode, either at the transformer or elsewhere outside the building.

Exception: **The additional grounding connection shall not be made on high-impedance grounded neutral systems. The system shall meet the requirements of Section 250-36.**

(3) Dual Fed Services. For services that are dual fed (double ended) in a common enclosure or grouped together in separate enclosures and employing a secondary tie, a single grounding electrode connection to the tie point of the grounded circuit conductor from each power source shall be permitted.

Exception No. 5: Where the main bonding jumper specified in Sections 250-23(b) and 250-79 is a wire or busbar, and is installed from the neutral bar or bus to the equipment grounding terminal bar or bus in the service equipment, the grounding electrode conductor shall be permitted to be connected to the equipment grounding terminal bar or bus to which the main bonding jumper is connected.

(Paragraph text) A grounding connection shall not be made to any grounded circuit conductor on the load side of the service disconnecting means.

Exception No. 1: A grounding electrode conductor shall be connected to the grounded conductor of a separately derived system in accordance with provisions of Section 250-26(b).
Exception No. 2: A grounding conductor connection shall be made at each separate building where required by Section 250-24.
Exception No. 3: For ranges, counter-mounted cooking units, wall mounted ovens, clothes dryers, and meter enclosures as permitted by Section 250-61.

(4) Main Bonding Jumper as Wire or Bus Bar. Where the main bonding jumper specified in Section *250-28* is a wire or busbar and is installed to the equipment grounding terminal bar or bus in the service equipment, the grounding electrode conductor shall be permitted to be connected to the equipment grounding terminal bar or bus to which the main bonding jumper is connected.

(5) Load-Side Grounding Connections. A grounding connection shall not be made to any grounded circuit conductor on the load side of the service disconnecting means *except as otherwise permitted in this article.*

(FPN) See Section *250-30(b)* for separately derived system, *Section 250-32* for connections at separate building or structures, and *Section 250-142* for use of the grounded circuit conductor for grounding equipment

(Three exceptions are combined into one Fine Print Note.)

Revision Keys

Special attention has been given to wording and making statements in full sentences. Many previous exceptions are just reworded into a complete sentence.

110-3(b) Installation and Use. Listed and labeled equipment shall be installed and used, ~~or both~~, in accordance with any instructions included in the listing or labeling.

Cross Check

1996	1999
210-52(c)(5) Receptacle Location. Receptacle outlets shall be located not more than 18 in. (458 cm) above the countertop.	**210-52(c)(5) Receptacle Location.** Receptacle outlets shall be located *above*, but not more than 18 in. (458 cm) above the countertop

Revision Keys

Two things happen to the Fine Print Notes (FPN). If the FPN is, or adds to, the intent of the rule, it is deleted as an FPN and made a part of the text. Example:

Cross Check

1996	1999
240-6 Standard Ampere Ratings 　**(a) Fuses and Fixed Trip Circuit** 　**Breakers** (FPN) It is not the intent to prohibit the use of nonstandard ampere ratings for fuses and inverse time circuit breakers.	**240-6 Standard Ampere Ratings** 　**(a) Fuses and Fixed Trip Circuit** 　**Breakers** The use of fuses and inverse time circuit breakers with nonstandard ampere ratings shall be permitted.

Many of the items used to refer to another part of the *Code* are deleted, because the *Code* should not be considered as a reference manual. Cross references should be in the Index. For example, in Article 100, the following are deleted.

~~General-Use Snap Switch: See under "Switches."~~
~~General-Use Switch: See under "Switches"~~

Revision Key

In several articles, the section referencing Article 250 Grounding have been deleted. Example: ~~300-9 Grounding Metal Enclosures.~~ This section required the grounding of metal raceways, boxes, cabinets, and fittings as required by Article 250. These sections are considered redundant as the grounding requirements are covered in Article 250.

ARTICLE 90
INTRODUCTION

■ ■

90-2(a)(4) Installation of optical fiber cables and raceways are now covered by the *Code*.

90-6 Formal Interpretations

Move the FPN following 90-6 Formal Interpretations and combine it into the text. It indicates where to find the established rules for an interpretation.

90-9 Metric Units of Measurement

Delete the FPN following 90-9 referencing location of metric measurements.

Reason

The conversions are already listed throughout the *Code*.

NEW TERM

The new term *"Permissive Rule"* applies to rules that identify actions that are allowed but not required. Many variations of this new term are used when exceptions are moved into the main body of a section.

RULE USE   *REVISION—NEW*
FPN   *NEW*

INTRODUCTION
TYPES OF RULES AND MATERIALS

90-5 (a)	Mandatory Rule	SHALL
		SHALL NOT
90-5 (b)	Permissive Rule	PERMITTED
	Conditional	NOT REQUIRED
		UNLESS
90-5 (c)	Explanatory Material	NOT ENFORCEABLE

90-5

REVISION

The section is revised into a subsection setting out each as part (a), (b), and (c). The one new term is *"Permissive Rules." Permissive rules identify actions that are allowed but not required. Permissive rules describe options or alternate methods and are characterized by terms "shall be permitted" or "shall not be required." The word "may" is not used to indicate a permissive rule.*

Reason

The new subsections highlight the different terms. With the introduction of the term "Permissive Rules," many of the exceptions in the *Code* are now restated as a positive statement in the text.

Comment

Two changes make the *Code* more User-Friendly. A long paragraph is outlined into subsections with the three main terms in boldface type. The introduction of the term "Permissive Rules" also clarifies options permitted. An understanding of this section will help recognize and understand the many *Code* changes. There is no change in the term "Mandatory." It is a firm positive statement with the use of the word SHALL. The following shows an exception that has been made a "Permissive Rule" in the text:

    110-34 Work Space and Guarding

    (c) Locked Rooms or Enclosures. The entrance to all buildings, rooms, or enclosures containing exposed live parts or exposed conductors operating at over 600 volts, nominal shall be kept locked *unless such entrances are under the observation of a qualified person at all times.*

90-5 NEW FPN

The format and language of the Code follow the NFPA guidelines as published in NEC® style manual. Manuals are available from NFPA.

Reason

The FPNs are not an enforceable part of the *Code.* This FPN has been added because some inspectors have tried to enforce some of the FPNs as part of the *Code.*

CHAPTER 1
GENERAL

■ ■

ARTICLE 100

DELETION

A multitude of cross references to other sections of the *Code* are deleted through-out *Article 100.* The following is an example of a cross-reference deletion:

Accessible (As Applied to Equipment) Admitting close approach: not guarded by locked doors, elevation, or other effective means. ~~(See "Accessible Readily").~~

~~**Appliance Branch Circuit:** See "Branch Circuit Appliance."~~

Festoon Lighting

The definition for festoon lighting is moved to *Article 100* from *Section 225-6(b),* Outside Branch Circuits and Feeders, with no change in wording.

Multioutlet Assembly

This definition is revised to include power poles.

Nonincendive Circuits

The definition for "Nonincendive Circuits" is revised and a new definition "Nonincendive Field Wiring" is also introduced.

Power Outlets.

The term *"park trailers"* is added to this definition.

Service: The conductors and equipment for delivering *electric* energy from the *serving utility* to the wiring system of premises served.

Reason

Other supply systems are normally separately derived systems.

Service Equipment

The necessary equipment, usually consisting of a circuit breaker*(s)* or switch*(es)* and fuse*(s),* and their accessories, *connected to the load end of the service conductors ~~located near the point of entrance of supply conductors~~* to a building or other structure, or an otherwise defined area, and intended to constitute the main control and means of cutoff of the supply.

Reason

The term "supply conductors" includes feeder and branch circuit conductors.

4

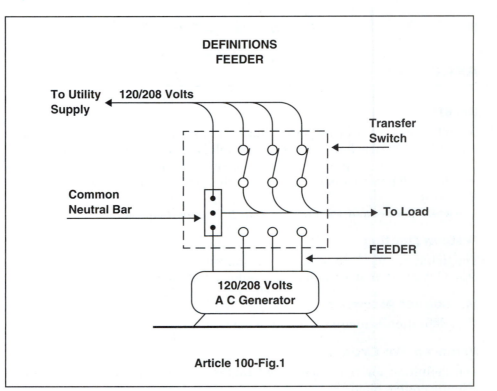

Article 100-Fig.1

REVISION

Feeder: All circuit conductors between the service equipment, the source of a separately derived system, *or other power supply* source and the final branch-circuit overcurrent device.

Reason

The previous definition did not cover the installation of an emergency system when the grounded conductor was common to the supply and to the generated emergency supply.

FESTOON LIGHTING

DEFINITIONS

ARTICLE 100

MOVED

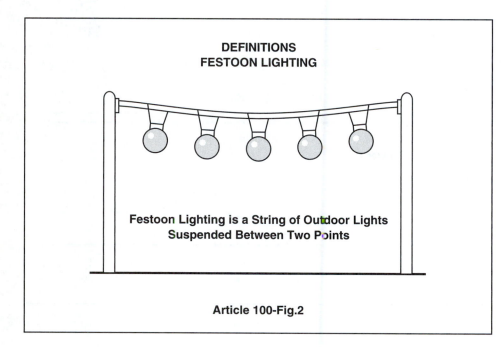

DEFINITIONS
FESTOON LIGHTING

Festoon Lighting is a String of Outdoor Lights
Suspended Between Two Points

Article 100-Fig.2

DEFINITION

Festoon lighting is a string of outdoor lights suspended between two points.

Reason
The term "festoon lighting" appears in more than two articles and is therefore properly put in Article 100. This conforms to the *NEC*® style manual.

Comment
The definition was moved from 225-6(b) with no change in the wording.

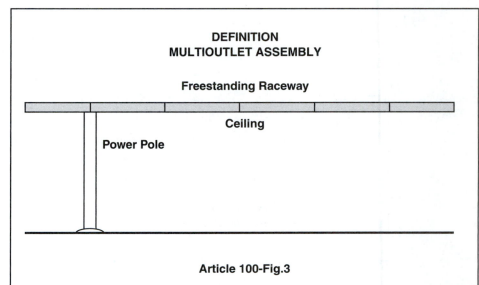

DEFINITION
MULTIOUTLET ASSEMBLY

Freestanding Raceway

Ceiling

Power Pole

Article 100-Fig.3

REVISION

Multioutlet Assembly. A type of surface, flush, or *freestanding* raceway designed to hold conductors and receptacles, assembled in the field or at the factory.

Reason
The revision is made so as to include power poles.

NONINCENDIVE CIRCUIT   *REVISED*

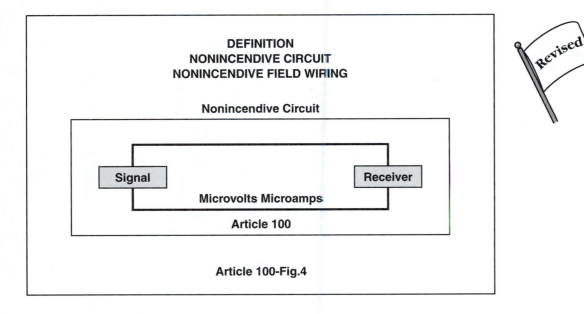

DEFINITION
NONINCENDIVE CIRCUIT
NONINCENDIVE FIELD WIRING

Nonincendive Circuit

Signal   Receiver

Microvolts Microamps

Article 100

Article 100-Fig.4

REVISION

NONINCENDIVE CIRCUIT

A circuit, **other than field wiring**, in which any arc or thermal effect produced, under intended operating conditions of the equipment, is not capable, under specified test conditions, of igniting the flammable gas-, vapor-, or dust-air mixture.

NONINCENDIVE FIELD WIRING   *NEW*

Wiring that enters or leaves an equipment enclosure and, under normal operating conditions of the equipment, is not capable, due to arcing or thermal effects, of igniting the flammable gas-, vapor-, or dust-air mixture. Normal operation includes opening, shorting, or grounding the field wiring.

Reason
These terms are used in more than one article of the *Code*. There was a need for the second definition because all nonincendive circuits contain nonincendive wiring.

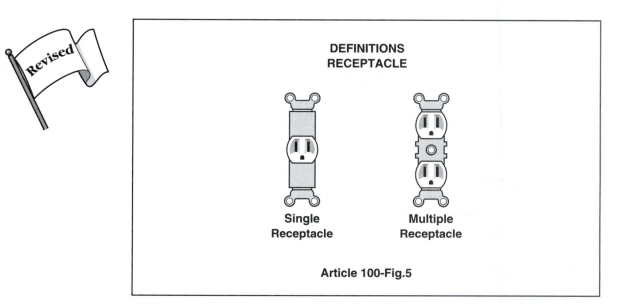

**DEFINITIONS
RECEPTACLE**

Single
Receptacle

Multiple
Receptacle

Article 100-Fig.5

REVISION

A **receptacle** is a contact device installed at the outlet for the connection of ~~a single contact device~~ ***an attachment plug.*** A **single receptacle** is a single contact device with no other contact device on the same yoke. A **multiple receptacle** is ~~a single device consisting of~~ two or more contact devices on same yoke.

Reason

When a multiwire branch circuit supplied more than one device or piece of equipment on the same yoke, the 1996 *Code, Section 210-4(b)* required a disconnecting means for each device.

Comment

As the 1996 *Code* was revised, a lot of the FPNs in *Article 100* were woven into the text, bringing this particular definition into conflict with *Section 210-4(b).*

ARTICLE 110 GENERAL

Article 110 Rearranged

Cross Check

1996	1999
Part A General Sections 110-2 through 110-22	Part A General Sections 110-2 through 110-22

New
    Section 110-16
    Section 110-17
	Part B 600 Volts Nominal and Under
	Section 110-26
	Section 110-27

Part B Over 600 Volts Nominal
    Sections 110-30 through 110-40
	Part C Over 600 Volts Nominal
	    Sections 110-30 Through 110-40

Section 710 Part F Tunnel Installations Sections 710-51 through 710-59	Part D Tunnel Installations, Over 600 Volts Nominal Sections 110-51 Through 110-59

110-10 Circuit Impedance and Other Characteristics

The term "~~withstand~~" is replaced with the word *"current"* and this new sentence. *Listed products applied in accordance with their listing shall be considered to meet the requirements of this section.*

Reason
Withstand ratings are not marked on the product. Short-Circuit Current ratings are marked on the products.

110-11 Deteriorating Agents

Revise last sentence to read: *Equipment ~~approved for use in~~ identified only as "dry locations," "Type I." or "indoor use only"* shall be protected against permanent damage from the weather during building construction.

Reason
Electrical equipment is identified by listing or labeling by the manufacturer as to the environmental use it is tested for.

110-14(a) Terminals

The exception for No. 10 conductors and upturned lugs is deleted and rephrased as a positive statement in the text.

110-14(c) Temperature Limitations

All exceptions are deleted and rephrased into positive language in the text. The subsection is divided further into subdivisions so that locating and interpreting rules is quick.

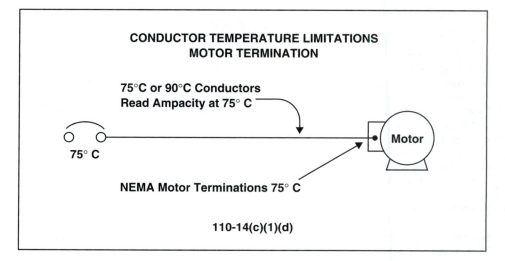

**CONDUCTOR TEMPERATURE LIMITATIONS
MOTOR TERMINATION**

75°C or 90°C Conductors
Read Ampacity at 75° C

75° C

Motor

NEMA Motor Terminations 75° C

110-14(c)(1)(d)

NEW RULE

Conductors having a 75°C or higher insulation rating are permitted for motors marked with Design letters B, C, D, or E, provided the ampacity of such conductors does not exceed the 75°C.

Reason

Motors with ratings of less than 100 amperes and designed according to NEMA standards are tested with 75°C conductors.

Comment

This indicates that 75°C or 90°C conductors are permitted for motor circuits provided the ampacity of the conductors is read under the 75°C column.

WORKING SPACE   110-26

REARRANGEMENT   *REVISED*

OUTLINE COMPARISON
Part B. 600 Volts, Nominal, or Less

1996	1999

110-16 Working Space About Electrical Equipment (600 Volts, Nominal, or Less)
(a) Working Clearances.
  Exception No. 1
  Exception No. 2
  Exception No. 3
Table 110-16(a)
  Condition 1
  Condition 2
  Condition 3
	110-26 Working Space About Electrical Equipment.
	
	(a) Work Space
	  (1) Depth of Working Space
	Table 110-26(a)
	  Condition 1
	  Condition 2
	  Condition 3
	  Exception No. 1
	  Exception No. 2
	  Exception No. 3
	  (2) Width of Working Space
	  (3) Height of Working Space

(b) Clear Space (c) Access and Entrance to Working (d) Illumination (e) Headroom	(b) Clear Space (c) Access and Entrance to Working (d) Illumination (e) Headroom

384-4 Installation	(f) Dedicated Equipment Space
	  Exception
	  (1) Indoors
	    (a) Dedicated Electrical Space
	    (b) Foriegn Systems
	    (c) Sprinkler Protection
	    (d) Suspended Ceilings
	  (2) Outdoors

110-26

REVISION

Several rules and statements were made in Section 110-26(a). The revision dismantles Section 110-16(a) of the 1996 *Code* and rearranges it into a more friendly format. The word "Clearances" is changed to ***"Spaces."***

Move the sentence "*Concrete, brick, or tile walls shall be considered as grounded*" from the first paragraph of the 1996 *Code* to the last sentence in Condition 2 of the 1999 *Code*.

Reason

The width, depth, and height rules were in one continuous section and hard to sort out. In the revision outline, the various width, depth, and height rules are now marked in bold type so they can be located easily. The word "clearances" is changed to "spaces," because that term is used throughout the section. The sentence was moved because it applies more directly to Condition 2.

Comment

This is one of many complex sections that have been rearranged with bold type and subheadings, making it easier to find the application of a particular rule. All the regulations under this section in the 1996 *Code* are in the 1999 *Code* with additional information.

EQUIPMENT SPACE

LOCKED DOORS

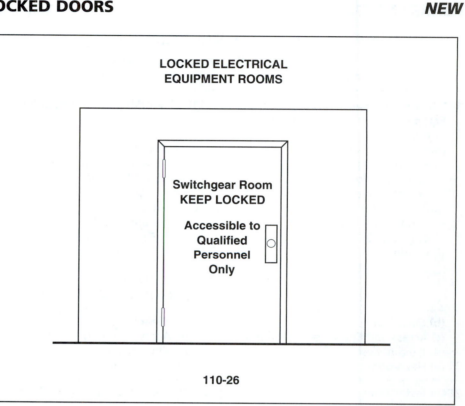

LOCKED ELECTRICAL EQUIPMENT ROOMS

Switchgear Room
KEEP LOCKED

Accessible to
Qualified
Personnel
Only

110-26

NEW RULE

Add as last sentence to Section 110-26. *Enclosures housing electrical appa-ratus that are controlled by lock and key shall be considered accessible to qualified persons.*

Reason

This rule was stated in Section 110-31 for Over 600 Volts, Nominal and is now also applicable to 600 Volts, Nominal or Under. It clarifies the fact that locked doors are considered accessible to qualified persons.

EQUIPMENT SPACE

FRONT AND BACK

110-26(a)(1) Exception

REVISED

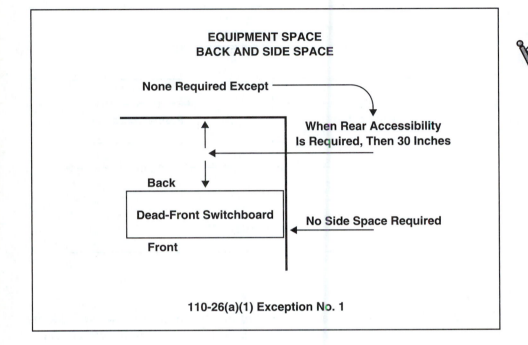

EQUIPMENT SPACE
BACK AND SIDE SPACE

None Required Except

When Rear Accessibility
Is Required, Then 30 Inches

Back

Dead-Front Switchboard

No Side Space Required

Front

110-26(a)(1) Exception No. 1

REVISION

The exception adds the word ***"sides"*** to the first sentence. This indicates that no working space is required in back of or on the sides of dead-front switchboards or motor control centers. The second sentence remains unchanged.

Reason

The side issue as well as the back issue need to be addressed.

Comment

The addition of the word "sides" in the first sentence indicates that no working space is required on the sides of the switchboard or motor control center. However, the second sentence is not changed to include "sides." Therefore, the 30 inches of working space is only required in back of the equipment when needed. No consideration is given to working space at the side of equipment.

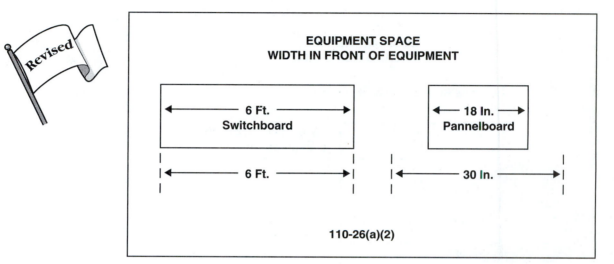

EQUIPMENT SPACE
WIDTH IN FRONT OF EQUIPMENT

6 Ft.
Switchboard

18 In.
Pannelboard

6 Ft.

30 In.

110-26(a)(2)

REVISION

The *width of the* working space *in front of electrical equipment* shall *be the width of the equipment or ~~not be less than~~* 30 in. (762 mm), **whichever is greater.** ~~wide in front of the electrical equipment~~

Reason

This revision clarifies the space requirement in front of electrical equipment. Working space of 30 inches is required, even when the panel board is less than that.

EQUIPMENT SPACE   110-26(a)(3)

HEIGHT OF EQUIPMENT   *REVISED*

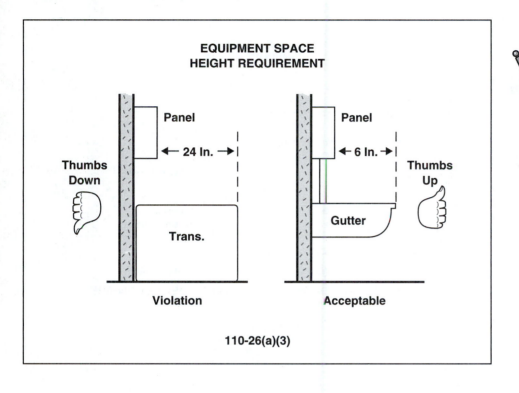

EQUIPMENT SPACE
HEIGHT REQUIREMENT

Panel

← 24 In. →

Thumbs
Down

Trans.

Violation

Panel

← 6 In. →

Thumbs
Up

Gutter

Acceptable

110-26(a)(3)

REVISION

1996 *Code:* ~~Equipment of equal depth shall be permitted.~~ 1999 *Code:* Within the height requirements of this section *other electrical installations located above or below electrical equipment are permitted to extend a maximum of 6 inches.*

Reason

The revision clarifies the rule without sacrificing its intent.

Comment

The 1996 rule was confusing and impractical. The intent of the rule is to keep electrical equipment, such as transformers, from being mounted below a panel which resulted in a serious problem of trying to reach over the transformer in order to work on the panel. However, when everything was aligned, it resulted in requiring a 4-inch-deep panel, mounted above a 6-inch auxiliary gutter to be spaced out a distance equal to the 6-inch auxiliary gutter.

EQUIPMENT SPACE 110-26(f)

DEDICATED SPACE *MOVED AND REVISED*

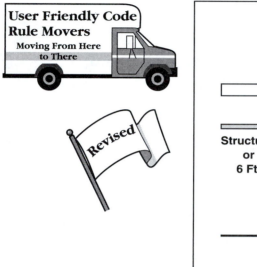

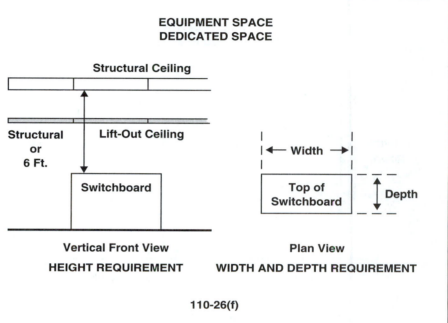

MOVE

Section 384-4 has been relocated to Section 110-26(f). Dedicated space is 6 ft. above equipment or structural ceiling, whichever is lower. It still applies to the switchboards and panelboards of Article 384.

Reason

This move gathers all spacing requirements for electrical equipment in a common place. The move was made by one of the "User-Friendly" task groups.

Comment

Section 110-16(f) covers the space dedicated to the installation of electrical equipment. In addition to the illustrated dedicated space, Section 110-26(a) must also be complied with. The following illustrates how the move and change in section numbers is made.

Cross Check

1996	1999
Article 384 Switchboard and Panelboards	110-26 Space About Electrical Equipment (600 Volts, Nominal or Less)
384-4 Installation	(f) Dedicated Equipment Space
(a) Indoors	(1) Indoors
(1) Width and Depth Part	(a) Dedicated Electrical Space
(2) Working Clearances Part	
Exception Part of exception	*Exception*
(1) Width and Depth Part	(b) Foreign Systems
(1) Width and Depth Part	(c) Sprinkler Protection
(1) Width and Depth Part	(d) Suspended Ceiling
(b) Outdoors	(2) Outdoors

CHAPTER 1 PART C. OVER 600 VOLTS, NOMINAL

110-31(c) ~~Metal~~ Enclosed Equipment Accessible to Unqualified Persons

The word "metal" has been deleted from this section, and the word "nonmetallic" has been added. Both nonmetallic and metallic enclosures for outdoor switchgear that are accessible to the general public must now be protected against vandalism by making all exposed nuts and bolts not readily removable.

The *Exception* following Section 110-31(c) has been deleted, reworded, and made a part of the text with no change in intent.

110-34 Working Space

Insert at the beginning of section *Except as elsewhere required or permitted in this Code.* No change in basic rule.

110-34(c) Locked Rooms and Enclosures

The *Exception* following Section 110-34(c) has been deleted, reworded, and made a part of the text with no change in intent.

Change in last sentence "Where the voltage exceed 600 volts, nominal," permanent and conspicuous warning signs shall be provided, ~~substantially~~ *reading as follows:*

DANGER—HIGH VOLTAGE—KEEP OUT

Section 710-9 Protection of Service Equipment, Metal-Enclosed Power Switchgear, and Control Assemblies

This section is moved with no change in title or rule and is now *110-34(f).*

Section 710-32 Circuit Conductors

This section is moved with no change in title or rule and is now coordinated with and into *Section 110-36.*

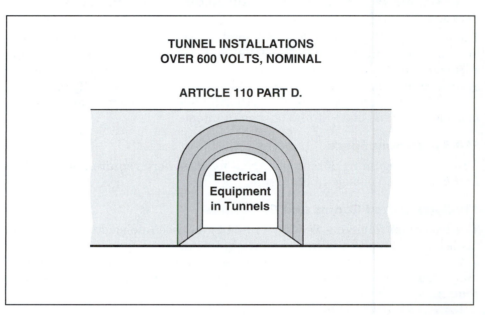

TUNNEL INSTALLATIONS
OVER 600 VOLTS, NOMINAL

ARTICLE 110 PART D.

Electrical Equipment in Tunnels

MOVE

Section F. of Article 710, Tunnel Installations, is moved, without change in the intent of the rules, from *Article 700* to a new **Part D. of Article 110** entitled **Tunnel Installations, Over 600 Volts, Nominal**.

Reason

Article 710, Over 600 Volt,s Nominal, has been deleted from the 1999 *Code*. A new **Article 490, Equipment Over 600 Volts Nominal,** will take its place.

Comment

The following is a numbers cross check of the moved sections.

Cross Check

1996	1999
710 Part F. Tunnel Installations	**110 Part D. Tunnel Installations, Over 600 Volts Nominal**

710-51 General
  (a) Covered
  (b) Other Articles
  (c) Protection Against Physical Damage
	110-51 General
	  (a) Covered
	  (b) Other Articles
	  (c) Protection Against Physical Damage

710-52 Overcurrent Protection
710-53 Conductors
710-54 Bonding and Equipment Grounding Conductor
  (a) Grounded and Bonded
  (b) Equipment Grounding Conductor
	110-52 Overcurrent Protection
	110-53 Conductors
	110-54 Bonding and Equipment Grounding Conductor
	  (a) Grounded and Bonded
	  (b) Equipment Grounding Conductor

710-55 Transformers, Switches, and Electrical Equipment **710-56 Energized Parts** **710-57 Ventilation System Controls**	**110-55 Transformers, Switches, and Electrical Equipment** **110-56 Energized Parts** **110-57 Ventilation System Controls**
710-58 Disconnecting Means **710-59 Enclosures** ~~**710-60 Grounding**~~	**110-58 Disconnecting Means** **110-59 Enclosures**

Section 710-60 is deleted because it is covered in *Article 250.*

CHAPTER 2
WIRING AND PROTECTION

■ ■

ARTICLE 200

The revision of Article 200 consists of some changes, rearrangement, and rewording for clarity. Some of the exceptions are deleted or reworded and subdivided into the text.

DELETION 200-10(b) Receptacle Plugs and Connectors

~~Exception: Terminal identification shall not be required for 2-wire non-polarized attachment plugs.~~

Reason

The exception is deleted because it is not covered by the main rule, which is directed at polarized receptacles.

The following is a cross check of the rearrangement of Article 200.

Cross Check

1996	1999
200-1 Scope (FPN) **200-2 General** *Exception*	**200-1 Scope** (FPN) **200-2 General** Part of text
200-3 Connection to Grounded System **200-6 Means of Identifying Grounded Conductors** **(a) Size No. 6 and Smaller** **200-6(a)** *Exception No. 4* *Exception No. 5* *Exception No. 2* Part of **200-6(a)** **(b) Sizes Larger Than No. 6** **(c) Flexible Cords** **(d) Grounded Conductors of Different Systems** New **200-6(b)** *Exception* **200-6(a)** *Exception No. 1*	**200-3 Connection to Grounded System** **200-6 Means of Identifying Grounded Conductor** **(a) Size No. 6 and Smaller** Also permitted (1) (2) (3) (4) **(b) Sizes Larger Than No. 6** **(c) Flexible Cords** **(d) Grounded Conductors of Different Systems** **(e) Grounded Conductors of Multiconductor Cables** *Exception No. 1* *Exception No. 2*
200-7 Use of White or Natural Gray Color **200-7** *Exception No. 4* **200-7** *Exception No. 1* *Exception No. 2* *Exception No. 3*	**200-7 Use of Insulation of White or Natural Gray Color or Three Continuous White stripes.** **(a) General** (1) White Natural Gray (2) Three Continuous White Stripes (3) Mark at Terminals **(b) Circuits of Less Than 50 Volts** **(c) Circuits of 50 Volts or More** (1) (2) (3)
200-9 Means of Identification of Terminals *Exception*	**200-9 Means of Identification of Terminals** *Exception*
200-10 Identification of Terminals **(a) Device Terminals** *Exception No. 1* *Exception No. 2* *Exception No. 3* **(b) Receptacles, Plugs, and Connectors** *Exception* Deleted (FPN) **(c) Screw Shells** **(d) Screw-Shell Devices with Leads** **(e) Appliances** **200-11 Polarity of Connections**	**200-10 Identification of Terminals** **(a) Device Terminals** Part of text Part of text *Exception* **(b) Receptacles, Plugs, and Connectors** (FPN) **(c) Screw Shells** **(d) Screw-Shell Devices with Leads** **(e) Appliances** **200-11 Polarity of Connections**

IDENTIFYING GROUNDED CONDUCTOR 200-6(a)

NO. 6 AND SMALLER *NEW*

New Arrival

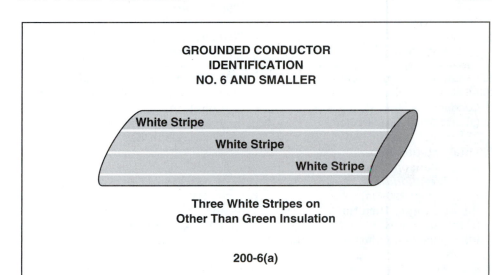

GROUNDED CONDUCTOR
IDENTIFICATION
NO. 6 AND SMALLER

White Stripe

White Stripe

White Stripe

Three White Stripes on
Other Than Green Insulation

200-6(a)

NEW IDENTIFICATION

The grounded conductor is permitted to be identified by:
1. White
2. Natural Gray
3. *three white strips on other than green insulation*
4. Other means.

Reason

This revision recognizes a product that is on the market.

Comment

The three white strips on the conductor are not specifically spaced. It is expected they will be 120 degrees apart.

IDENTIFYING GROUNDED CONDUCTOR 200-6(b)

LARGER THAN NO. 6 *REVISED*

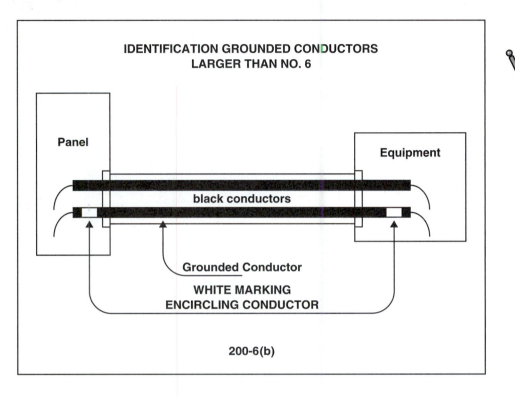

IDENTIFICATION GROUNDED CONDUCTORS
LARGER THAN NO. 6

Panel

Equipment

black conductors

Grounded Conductor

WHITE MARKING
ENCIRCLING CONDUCTOR

200-6(b)

Comment
A single insulated grounded conductor larger than No. 6 is required to be identified by a continuous white or natural gray finish *or three continuous white stripes on other than green insulation or* distinctive markings at terminations.

REVISION
The revision adds the following sentence: *This marking shall encircle the conductor or insulation.*

Reason
When the marking is only on one side of the conductor, it can be hidden from view, thereby creating an electrical hazard.

Comment
This revision demonstrates that common sense has to be written into the *Code.* Note that this only applies to conductors larger than No. 6.

USE OF WHITE INSULATED CONDUCTOR 200-7(c)(1)

CIRCUITS OF 50 VOLTS OR MORE
IN CABLE ASSEMBLIES

REVISED

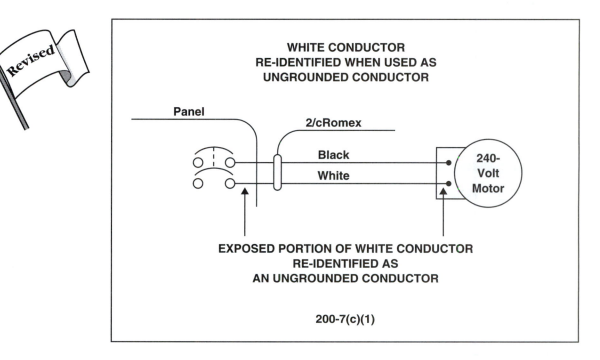

WHITE CONDUCTOR
RE-IDENTIFIED WHEN USED AS
UNGROUNDED CONDUCTOR

Panel

2/cRomex

Black

White

240-Volt Motor

EXPOSED PORTION OF WHITE CONDUCTOR
RE-IDENTIFIED AS
AN UNGROUNDED CONDUCTOR

200-7(c)(1)

REVISION

Revise *Exception No. 1* into a positive statement, add it to the text, and add the phrase, *"where part of a cable assembly."*

The white conductor can be used as an ungrounded conductor *where part of a cable assembly* and when permanently re-identified to indicate its use as an ungrounded conductor, by painting or other effective means at the termination and at each location where the conductor is visible and accessible.

Reason

When a white or natural gray conductor in a cable assembly, such as Romex or BX, is used for an ungrounded conductor and not marked, it creates an electrical hazard. Accidents have occurred on remodeling work where the white or natural gray conductor is installed as an ungrounded conductor and not identified.

Comment

This new exception is evidently directed at some flagrant *Code* violations, such as the use of two-conductor cable with a black and a white conductor supplying a single-phase 240-volt motor. In such an installation, the white conductor must be re-identified with the re-identification circling the white conductor.

USE OF WHITE INSULATED CONDUCTOR   200-7(c)(2)

CIRCUITS OF 50 VOLTS OR MORE
SWITCH SUPPLY   *NEW*

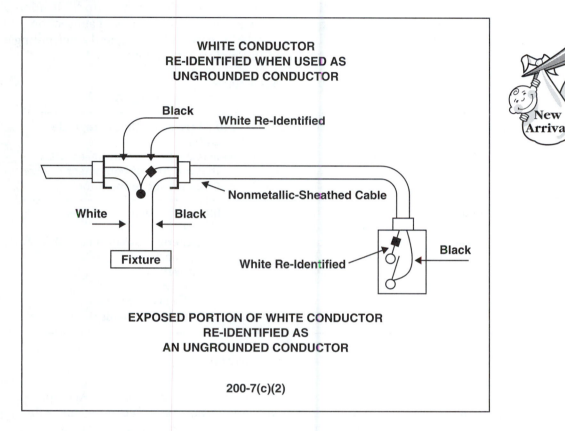

WHITE CONDUCTOR
RE-IDENTIFIED WHEN USED AS
UNGROUNDED CONDUCTOR

Black

White Re-Identified

Nonmetallic-Sheathed Cable

White   Black

Fixture   White Re-Identified   Black

EXPOSED PORTION OF WHITE CONDUCTOR
RE-IDENTIFIED AS
AN UNGROUNDED CONDUCTOR

200-7(c)(2)

NEW RULE

Where a cable assembly with a white or natural gray conductor is used for a switch leg or as switch loops for 3-way or 4-way switches, it is required that:
1. It is not to be used as the return conductor.
2. *It shall be permanently re-identified to indicate its use by painting or other effective means at each location where the white, natural gray, or three continuous white strip conductor is visible and accessible.*

Reason

This revision guards against the misapplication of the white or natural gray insulated conductor.

Comment

This revision now means that when the white conductor in nonmetallic-sheathed cable is used for a switch loop or as travelers for 3-way or 4-way switches, the white conductor is required to be re-identified as other than a grounded conductor at all switch locations, pull junction boxes, and lighting outlets.

ARTICLE 210

210-52 Dwelling Unit Receptacle Outlets

250-52(a) This long, complicated paragraph containing a multitude of information has been broken into "subsections" and "subdivisions of subsections." Each breakdown is given a boldface title for ease of locating and interpreting the rule. One new rule is introduced in Section 210-52 text.

Cross Check

1996	1999
210-52 Dwelling Unit Receptacle Outlets	210-52 Dwelling Unit Receptacle Outlets
(a) General Provisions Part of text	New text—Don't count receptacles that are part of equipment or over 51/2 feet above the floor
250-52(a) *Exception* FPN applicable to exception **(a) General Provisions** Part of (a) Part of text Part of text Part of text Part of text Part of text	Included in text FPN **(a) General Provisions** Part of (a) (1) Spacing (2) Wall Space (a) 2 ft (b) Fixed panels (c) Room dividers

210-52 Dwelling Unit Receptacle Outlets

210-52(a)(2)(a) Wall Space.
This section is revised to include measurements around corners. Any space 2 feet or more in width, ***including space measured around corners,*** and unbroken along the floor line by doorways, fireplaces, and similar openings.

(b)Small Appliances, subdivision *(3)* is directed at kitchen countertop receptacles. A new sentence is added: ***No small appliance branch circuit shall serve more than one kitchen.***

210-52(c)(5) Countertop Receptacle Located Below Countertop

The exception permitting receptacles to be mounted below the countertop is revised by deleting, rewording, and adding extra information. Exception: *In order to comply with a and b below, receptacles are permitted to be mounted not more than 12 inches below the countertop provided the countertop does not extend more than 6 inches beyond its support base.*
a. Constructed for physically impaired.
b. *On island and peninsular countertops where the countertop is flat across its entire surface (no back splashes, dividers, or other raised features), and there is no means to mount a receptacle within 18 inches (458 mm) above the countertop, such as in an overhead cabinet.*

Reason
This change relieves the AHJ of the decision-making responsibility. The revision makes the rule more uniform by indicating constructions that preclude mounting a receptacle above the countertop. The previous rule seems to have been interpreted in a variety of ways.

210-60 Guest Rooms

This short section is revised and outlined, and has new rules added.

Cross Check

1996	1999
210-60 Guest Rooms	**210-60 Guest Rooms**
Text	**(a) General**
	(b) Receptacle placement
Exception	Included in text
New	2 outlets readily accessible
New	Guards for underbed receptacles

Reason

This section needed clarification. Two receptacles should be sufficient for the operation of tenant's electronic equipment. Receptacles behind the bed should be protected. When a receptacle is installed behind the headboard of a built-in hotel bed, the receptacle, or cord plugged into the receptacle, are damaged and present a real electrical hazard.

210-70 Lighting Outlets Required

This section is rearranged with subheadings.

Cross Check

1996	1999
210-70 Lighting Outlets Required	**210-70 Lighting Outlets Required**
(a) Dwelling Units	**(a) Dwelling Units**
Part of text	**(1) Habitable Rooms**
Exception No. 1	*Exception No. 1*
Exception No. 3	*Exception No. 2*
Part of text	**(2) Additional Locations**
Exception No. 2	*Exception*
Part of text	**(3) Storage and Equipment Space**
(b) Guest Rooms	**(b) Guest Rooms**
(c) Other Locations	**(c) Other Locations**

Section 210-70(a)(1) has been revised to also require a ***wall switched-controlled lighting outlet in a bathroom*** even though a bathroom is not considered to be a habitable room.

Comment

The intent of the rules has not changed. The new format makes the information easier to find. Rewording clarifies the rule.

MULTIWIRE BRANCH CIRCUITS IDENTIFICATION
UNGROUNDED CONDUCTORS

120 / 240 VOLTS   **Common Box**

Brown

Purple   → To Receptacles

White Red Stripe

120–208 VOLTS

Red

Black

Blue   → To Lighting

White   Fixture

210-4(d)

REVISION

Where more than one nominal voltage system exists in a building, each ungrounded conductor *of multiwire branch circuit, where accessible,* must be identified by phase and system. Means of identification is required to be posted at each branch-circuit panel board. *Permitted means of identification are: separate color coding, marking tape, tagging, and other approved means.*

Reason

The rewording makes it clear that this rule applies to multiwire branch circuits only and not to the entire system. The identification is required to avoid any cross connections between systems.

Comment

The word "system" was sometimes interpreted as meaning that all feeder conductors and branch circuit conductors had to be identified when two ungrounded systems were used in a building. This rule is directed at the safety hazard in the branch circuit instructions under the heading "Multiwire Branch Circuits," and it only applies to multiwire branch circuits. The color code used in the illustration is not required; it shows only potential colors that might be used. The neutrals are identified as per 200-6(d).

BRANCH CIRCUITS CONDUCTOR
IDENTIFICATION

DELETED—REVISED

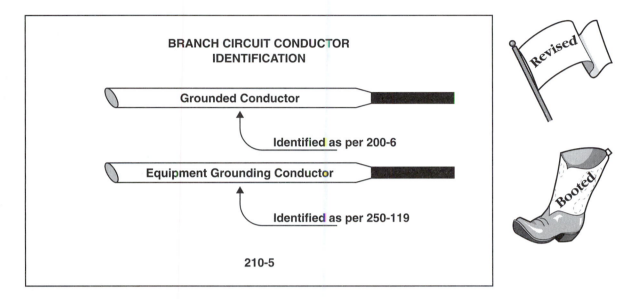

Comment

The title of Section 210-5 is revised from ~~Color Code for Branch Circuits~~ to *Identifications for Branch Circuits.*

DELETION—REVISION

210-5(a) Grounded Conductor. The entire subsection and the exception are deleted. A new single sentence is added: ***The grounded conductor of a branch circuit shall be identified in accordance with Section 200-6.***

Reason

The rule and two exceptions presented in this subsection in the 1996 *Code* duplicated the rules and exceptions presented in 200-6.

DELETION—REVISION

210-5(b) Equipment Grounding Conductor. The entire subsection and the exception are deleted. A new single sentence is added: ***The equipment grounding conductor of a branch circuit shall be identified in accordance with Section 250-119.***

Reason

The rule and exception presented in this subsection in the 1996 *Code* duplicated the rules and exceptions presented in Article 250.

Comment

By focusing on branch circuits only in Article 210, Branch Circuits, the 1999 *Code* is more user friendly. Section 200-6 requires the grounded conductor to be white or natural gray or three white strips on other than green insulations, and *Section 250-119* requires the equipment-grounding conductor to be green or green with one or more yellow strips.

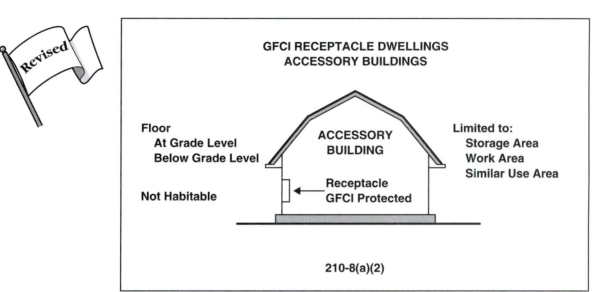

REVISION

GFCI receptacles must be installed in the accessory buildings *having a floor located at or below grade level, not intended as habitable rooms, and limited to* used for storage *areas*, work areas, *and areas of similar use.*

Reason

The terms "unfinished" and "grade-level portion" are open to a multitude of interpretations.

Comment

The *Code* does not require the installation of lights or receptacles in these buildings, but when they are installed, they must conform to the *Code*. This rule is directed at lights and/or receptacles installed in prefabricated buildings. One concern is the use of an outdoor power tool extension cord plugged into an accessory building receptacle. When the building is set on a dirt floor, slightly raised wood floor, or concrete floor, and the receptacles are installed at or below grade level, they must be GFCI protected.

GFCI RECEPTACLES—
OTHER THAN DWELLINGS

210-8(b)(2) Exception

ROOFTOP INSTALLATIONS

NEW

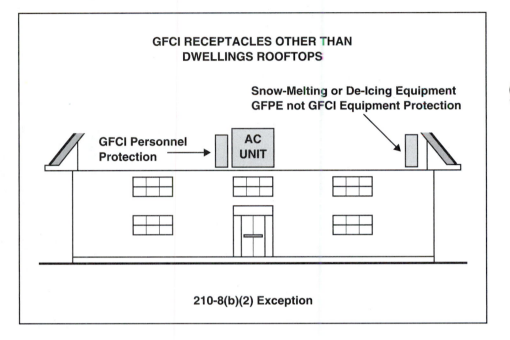

GFCI RECEPTACLES OTHER THAN
DWELLINGS ROOFTOPS

Snow-Melting or De-Icing Equipment
GFPE not GFCI Equipment Protection

GFCI Personnel
Protection

AC
UNIT

210-8(b)(2) Exception

New
Arrival

BASIC RULE

All 125-volt 15- or 20-amp receptacles installed on rooftops are required to be GFCI protected for personnel.

NEW EXCEPTION

Receptacles that are not readily accessible and are supplied by a dedicated branch circuit for electric snow-melting or de-icing equipment are permitted to be installed in accordance with the applicable provisions of Article 426.

Reason

The branch circuits used to supply snow-melting or de-icing equipment are required by Section 426-28 to have ground-fault protection for equipment (GFPE). This includes the receptacle connected in the branch circuit and used for the snow-melting equipment.

Comment

This equipment is permitted to be cord- and plug-connected by Section 426-3. The exception is necessary because a GFCI takes a higher current to operate for the protection of equipment than a GFCI does for personnel protection.

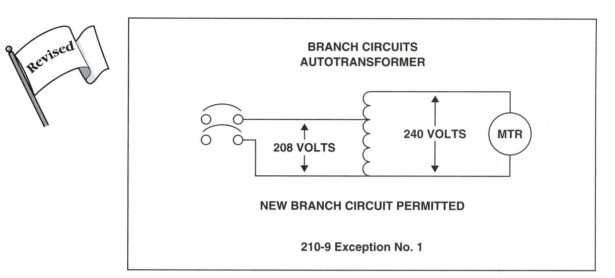

BASIC RULE

Branch circuits shall not be derived from autotransformers unless the grounded conductor is common to the primary and secondary of the autotransformer.

REVISION

*An autotransformer shall be permitted ~~for a to extend or add a branch circuit for an equipment load~~ without the connection to a ~~similar~~ grounded conductor ~~when~~ **where** transforming from a nominal 208-volt to a nominal 240-volt supply or similarly from 240 volts to 208 volts.*

Reason

If an autotransformer is safe to use on an existing or an added branch circuit, it should also be safe to use on new branch circuits.

Comment

This revision now permits the use of an autotransformer on new circuits as well as existing circuits without the use of a grounded conductor.

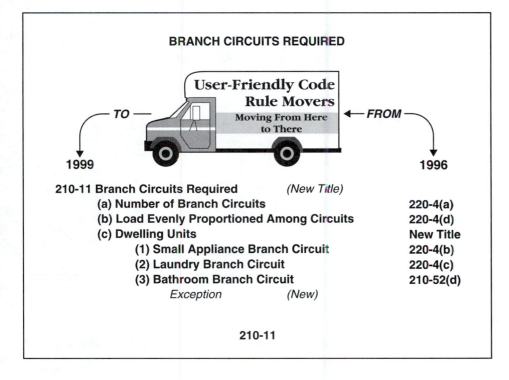

BRANCH CIRCUITS REQUIRED

TO   **1999**   *FROM*   **1996**

210-11 Branch Circuits Required	*(New Title)*	
(a) Number of Branch Circuits		**220-4(a)**
(b) Load Evenly Proportioned Among Circuits		**220-4(d)**
(c) Dwelling Units		**New Title**
(1) Small Appliance Branch Circuit		**220-4(b)**
(2) Laundry Branch Circuit		**220-4(c)**
(3) Bathroom Branch Circuit		**210-52(d)**
Exception	*(New)*	

210-11

MOVED

New section number, same title and same rules are now in Section ***210-11, Branch Circuits Required***. All of Section 220-4 is moved from Article 220 to Article 210 with some rearrangement and change in section number. The rules presented in Section 220-4 have been moved without altering their intent.

Reason

A special ad hoc committee worked on coordinating branch circuit requirements into Article 210 and rearranging some of the rules in Article 210 for ease of location.

Comment

A bullet in the margin of Section 220-4 indicates a deletion; however, the rule has been moved rather than deleted. It is still very much a part of the *Code.*

Although everything in Section 220-4 has been moved, the Section number is not retired as has been the custom in the past. It is reused when Section 210-22 is moved to this location.

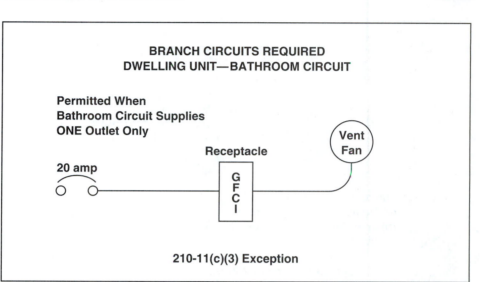

BASIC RULE

One 20-ampere branch circuit is required to supply bathroom receptacle outlet(s). No other outlets are permitted on this circuit.

NEW EXCEPTION

Where a 20-ampere branch circuit supplies a single outlet, other equipment within the same bathroom is permitted on this circuit in accordance with 210-23(a).

Reason

Frequently the GFCI-protected bathroom receptacle is also connected so as to protect the vent fan. The exception is needed to continue this practice, because the rule states "No other outlets are permitted on this circuit."

Comment

The basic rule requires a 20-ampere branch circuit for bathroom outlet(s), permitting this circuit to supply more than one receptacle in a bathroom or receptacles in two bathrooms. Note that the new exception reads: ***"branch circuit supplies a single outlet."*** Once a second outlet in the same bathroom or an outlet in another bathroom is connected to this 20-ampere bathroom circuit, the vent fan can no longer be connected to this circuit. Section 210-23(a), Permissible Loads on 20-Ampere Branch Circuits, permits the 20-ampere branch circuit to supply lighting outlets.

BRANCH CIRCUITS   210-12

ARC-FAULT CIRCUIT INTERRUPTER   *NEW*

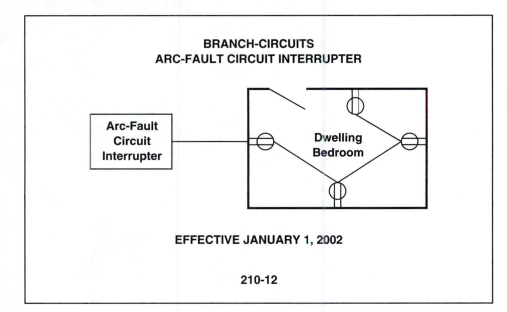

**BRANCH-CIRCUITS
ARC-FAULT CIRCUIT INTERRUPTER**

Arc-Fault
Circuit
Interrupter

Dwelling
Bedroom

EFFECTIVE JANUARY 1, 2002

210-12

New
Arrival

NEW RULE

An Arc-Fault Circuit Interrupter is a device intended to provide protection from the effects of arc faults by recognizing characteristics unique to arcing and by functioning to de-energize the circuit when an arc fault is detected.

Effective January 1, 2002, all branch-circuits that supply 125 volt, single phase, 15- and 20-ampere receptacle outlets installed in bedrooms of dwelling units will be required to have Arc-Fault Circuit Interrupter protection.

Reason

A large number of residential fires are said to be caused by an electrical arc. The arc-circuit interrupter is a device designed to protect against starting a fire when there is an unintended electrical arc on the circuit.

CONDUCTOR SIZING CONTINUOUS
AND NONCONTINUOUS LOADS   *MOVED AND REVISED*

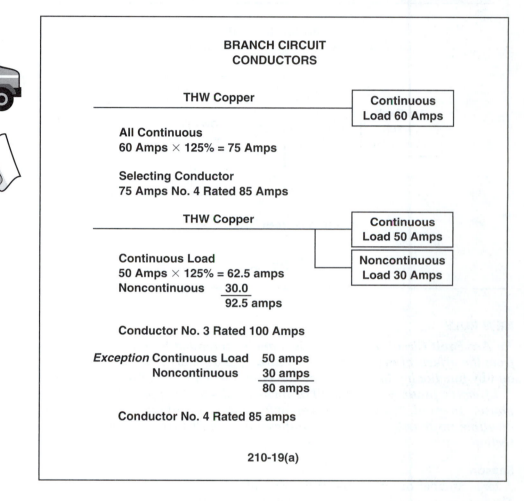

User Friendly Code
Rule Movers
Moving From Here
to There

Revised

**BRANCH CIRCUIT
CONDUCTORS**

THW Copper — Continuous Load 60 Amps

**All Continuous
60 Amps × 125% = 75 Amps**

**Selecting Conductor
75 Amps No. 4 Rated 85 Amps**

THW Copper — Continuous Load 50 Amps / Noncontinuous Load 30 Amps

**Continuous Load
50 Amps × 125% = 62.5 amps
Noncontinuous   30.0
                  92.5 amps**

Conductor No. 3 Rated 100 Amps

Exception **Continuous Load   50 amps
           Noncontinuous   30 amps
                              80 amps**

Conductor No. 4 Rated 85 amps

210-19(a)

MOVE

The requirements for sizing continuous and noncontinuous branch circuit conductors are moved from **210-22(c) to 210-19(a)**. The exception is also moved and revised.

Reason

The move is made with no change in the basic rule for conductors. In the 1996 *Code*, the branch circuit conductors and the branch circuit overcurrent protection for continuous loads were covered in Section 210-22(c). The 1999 *Code* deletes 210-22(c) and moves the rules applicable to conductors under the heading of conductors in Section 210-19. Overcurrent protection rules are moved under overcurrent protection in Section 210-20. This move separates rules into appropriate locations.

REVISION

The *Exception* that followed Section 210-22(c) of the 1996 *Code* is moved and reworded, and it now follows Section 210-19(a) of the 1999 *Code*.

Exception: When the assembly and the branch circuit overcurrent device is listed for 100 percent of their rating, **the ampacity of the branch circuit conductors shall be permitted to be not less than the sum of the continuous load plus the noncontinuous load.**

Reason

This clarification establishes a minimum for the conductor when the assembly and the overcurrent device are rated at 100 percent of their ampacity. This was not specifically noted in the previous *Code*.

BRANCH CIRCUITS   210-19(b)

CONDUCTORS MULTIOUTLET
BRANCH CIRCUITS   *MOVED AND REVISED*

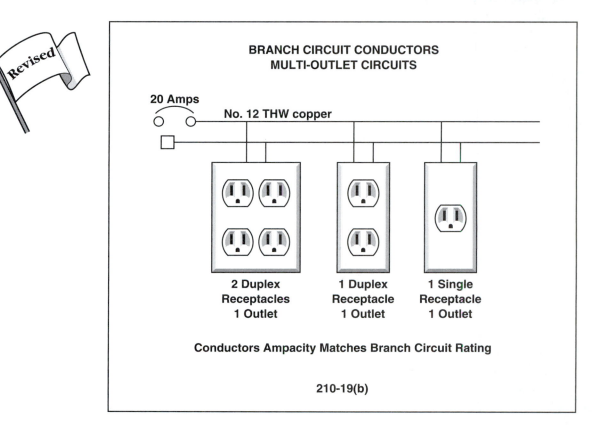

BRANCH CIRCUIT CONDUCTORS
MULTI-OUTLET CIRCUITS

20 Amps
No. 12 THW copper

2 Duplex
Receptacles
1 Outlet

1 Duplex
Receptacle
1 Outlet

1 Single
Receptacle
1 Outlet

Conductors Ampacity Matches Branch Circuit Rating

210-19(b)

MOVE
The rule is taken from 210-19(a) and moved to a new subsection 210-19(b).

REVISION
~~In addition,~~ Conductors ~~of multioutlet~~ **of** branch circuits supplying **more than one** receptacle for cord- and plug-connected portable loads shall have an ampacity of not less than the rating of the branch circuit. ~~Cable assemblies where the neutral conductor is smaller than the ungrounded conductors shall be so marked.~~

Reason
The revision makes it clear that the conductors are protected by the overcurrent device, where one outlet might supply more than one duplex receptacle for portable loads.

Comment
This new subsection pushes two subsections down. What was (b) is now (e), and what was (c) is now (d).

BRANCH CIRCUITS   210-20

OVERCURRENT PROTECTION   *REVISED*

**BRANCH CIRCUIT
OVERCURRENT PROTECTION**

1996	Cross Check	1999
210-20 Overcurrent Protection **210-22(c) Other Loads** *Exception*		**210-20 Overcurrent Protection** **(a) Continuous and Noncontinous Loads** *Exception*
Part of 210-20 **210-20** *Exception No. 1* *Exception No. 2*		**(b) Conductor Protection** *Exception No. 1* *Exception No. 2*
Part of 210-20		**(c) Equipment**
Part of 210-20		**(d) Outlet Devices**

210-20

REVISION
The revision brings a rule from 210-22(c) under the title Overcurrent Protection and breaks the several parts of 210-20(a) into subheadings.

Reason
The revision brings like things under the same heading and boldface subsection headings for ease of identification.

Comment
The move and revisions are made with no change in the intent of the rule.

OVERCURRENT PROTECTION CONTINUOUS AND NONCONTINUOUS LOADS *MOVED AND REVISED*

User Friendly Code Rule Movers
Moving From Here to There

Revised

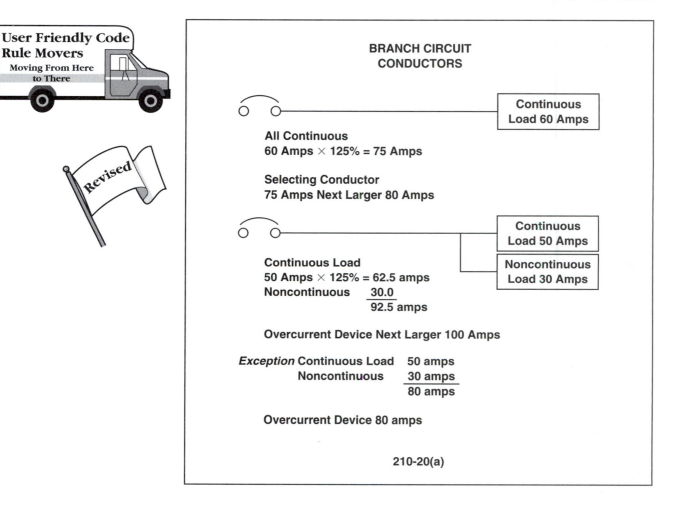

BRANCH CIRCUIT CONDUCTORS

Continuous Load 60 Amps

All Continuous
60 Amps × 125% = 75 Amps

Selecting Conductor
75 Amps Next Larger 80 Amps

Continuous Load 50 Amps

Noncontinuous Load 30 Amps

Continuous Load
50 Amps × 125% = 62.5 amps
Noncontinuous   30.0
                          92.5 amps

Overcurrent Device Next Larger 100 Amps

Exception Continuous Load   50 amps
              Noncontinuous   30 amps
                                        80 amps

Overcurrent Device 80 amps

210-20(a)

MOVE

The requirements for continuous and noncontinuous branch circuit overcurrent protection are moved from **210-22(c) to 210-20(a).** The exception is also moved and revised.

Reason

The rule is unchanged. In the 1996 *Code,* the rules for branch circuit conductors and the branch circuit overcurrent protection for continuous loads was covered in Section 210-22(c). The 1999 *Code* deletes 210-22(c) and moves the rules applicable to conductors to Section 210-19. Overcurrent protection rules are moved to Section 210-20. These moves put the rules in appropriate locations.

REVISION

The *Exception* following 210-22(c) in the 1996 *Code* is deleted, moved, and revised following 210-20(a) in the 1999 *Code.*

Exception: When the assembly and the branch circuit overcurrent device is listed for 100% of their rating, **the ampere rating of the overcurrent device shall be permitted to be not less than the sum of the continuous load plus the noncontinuous load.**

Reason

This clarification establishes a minimum for the conductor and the overcurrent device when the assembly and the overcurrent device are rated at 100% of their ampacity.

Caution

When the branch circuit overcurrent device is rated at 100% of its rating and there is no increase for the overcurrent device or the conductor, the protection of the conductor by the overcurrent device still must be considered. Example:

overcurrent device rated for 100%

overcurrent device rating = continuous load + noncontinuous load

overcurrent device = 100 + 30 = 130 amps

next larger overcurrent device: 150 amps

No. 4 THW copper is rated 130 amps. This is large enough for the total load, but it cannot be protected with a 150-ampere overcurrent device. The next larger conductor would be needed.

**BRANCH CIRCUITS
MAXIMUM LOADS**

MOVING

FROM	TO
1996	*1999*

210-22 ⟶ 220-4 Maximum Loads

210-22(a) ⟶ 220-4(a) Motor-Operated and Combination Loads

210-22(b) ⟶ 220-4(b) Inductive Loads

210-22(c) ⟶ 220-4(c) Range Loads

210-22

MOVED

Section 210-22, Maximum Loads, is moved to **220-4, *Maximum Loads***, creating a new section in Article 220 with no change in wording or intent.

Reason

This section is directed at "computed loads" and computed loads for branch circuits are covered in Part A of Article 220.

Comment

The new Section **210-11** was the result of moving branch circuit requirements from Article 220 to Article 210. Now the information relating to the computing of branch circuit loads is moved from Article 210 to Article 220 and a new **Section 220-4**.

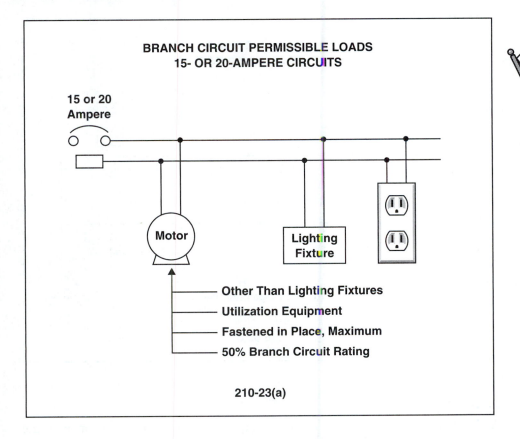

REVISION

The total rating of utilization equipment fastened in place, ***other than lighting fixtures,*** shall not exceed 50% of the branch-curcuit ampere rating where lighting units, cord- and plug-connected utilization equipment not fastened in place, or both are also supplied.

Reason

The revision makes clear that even though lighting fixtures are utilization equipment fastened in place, they are not the utilization equipment the 50% maximum is directed at.

Comment

For example, a 20-ampere general-purpose branch circuit could supply a utilization load, such as a motor, up to 50% of 20 amperes or 10 amperes. Receptacles, lighting fixtures, or cord- and plug-connected equipment could also be connected to the same 20-ampere branch circuit. What is connected for the remaining 10 amperes is limited by common sense.

BATHROOM RECEPTACLES

REVISED

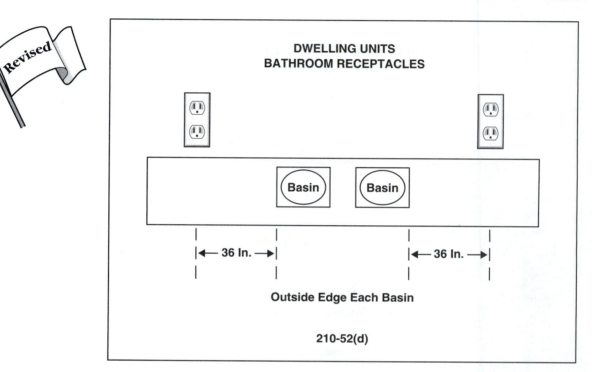

**DWELLING UNITS
BATHROOM RECEPTACLES**

|← 36 In. →| |← 36 In. →|

Outside Edge Each Basin

210-52(d)

REVISION

At least one wall receptacle outlet is required to be installed in a dwelling bathroom located *within 36 in. of the outside edge of the basin. This receptacle outlet is required to be located on a wall that is adjacent to the basin.*

Reason

The word "adjacent" was too flexible, resulting in the electrical contractor and the inspector making arbitrary measurements. The revised rule sets a definite measurement from a given point.

Comment

In some cases, the required receptacle was installed 4 or 5 feet from the basin, so an extension cord had to be used. This hazardous use of an extension cord should be eliminated. The contiguous wall location is required so that in narrow bathrooms (36 inches or less between basin and a wall) the receptacle will not be mounted behind the person using the basin.

UNFINISHED BASEMENT RECEPTACLES   *REVISED*

```
┌─────────────────────────────────────────────────────┐
│              DWELLING UNIT                            │
│       UNFINISHED BASEMENT RECEPTACLES                 │
│                                                      │
│            Dwelling Basement                         │
│   ┌──────────────────┬──────────────────────┐        │
│   │                  │                      │        │
│   │   Finished       │    Unfinished        │        │
│   │                  /                      │        │
│   │   Portion        │    Portion       ○   │        │
│   │                  │                      │        │
│   └──────────────────┴──────────────────────┘        │
│                                          ▲           │
│   GFCI RECEPTACLE REQUIRED ──────────────┘           │
│                                                      │
│                  210-52(g)                           │
└─────────────────────────────────────────────────────┘
```

REVISION

A receptacle must be installed in each basement *or unfinished basement area.*

Reason

The revision is made to clarify that a receptacle outlet is required in the unfinished portion of a basement when part of a basement is finished into habitable room and part is unfinished.

Comment

Section 210-8(a)(5) requires GFCI protection for receptacles installed in unfinished basements. Section 210-52(g) requires a receptacle in the basement. The previous wording permitted one receptacle in the finished portion of the basement and did not specifically cover the unfinished portion, thereby allowing an unprotected extension cord to be run into the unfinished basement.

ARTICLE 215

215-2 Minimum Rating and Size

(a) General. The rule requiring feeder conductors for continuous and noncontinuous loads to be 125% of the continuous load plus the noncontinuous load is moved from **Section 220-10(b)** to **215-2(a).**

Exception is also moved from **220-10(b)** with revision. The existing **(a)** and **(b)** have been re-identified as **(b)** and **(c)**, respectively, because of this move.

Add new subsection **(d) Individual Dwelling Unit and Mobile Home Conductors.** This is moved from Section **215-2**, though the basic rule remains. In addition, a new sentence is added: *Section 310-15(b)(6) shall be permitted to be used for conductor size.*

Section 310-15(b)(6) is the same as Note 3 to Table 310-16 of the 1996 *Code*, which allows special-size conductors for service-entrance conductors.

215-3 Overcurrent Protection

In addition to the existing sentence, the rule requiring feeder overcurrent protection for continuous and noncontinuous loads to be 125 percent of the continuous load plus the noncontinuous load is moved from Section **220-10(b)** to **215-3**.

The *Exception* is also moved from **220-10(b)** with some revision:

Where the assembly including the overcurrent devices protecting the feeder(s) are listed for operation at 100 percent of their rating the ampere rating of the overcurrent device shall be permitted to be not less than the sum of the continuous load plus the nonconinuous load.

Comment

This relocates rules for feeder conductors and feeder overcurrent protection under the more logical heading of feeders. This corresponds to the move made under branch circuits where continuous and noncontinuous loads are involved. The rearrangement and rewording of material is for better understanding and location of information.

Cross Check

1996	1999
215-2 Minimum Rating and Sizing (Part of 215-2 and 220-10(b)	**215-2 Minimum Rating and Sizing** (a) General
	Exception
220-10(b) *Exception*	**(b) Specific Circuits**
(a) Specific Circuits	(1)
	(2)
	(3)
	(4)
(b) Ampacity Relative to Service-Entrance Conductors Part of 215-2	**(b) Ampacity Relative to Service-Entrance Conductors**
	(c) Individual Dwelling Unit or Mobile Home Conductors
FPN No. 1	FPN No. 1
FPN No. 2	FPN No. 2
FPN No. 3	FPN No. 3
215-3 Overcurrent Protection Part of 220-3(b)	**215-3 Overcurrent Protection** Included in text
220-3(b) *Exception*	*Exception*

215-10 Ground-Fault Protection of Equipment (GFPE)

The basic rule requires GFPE for each feeder disconnect

on a solidly grounded wye system

rated 1000 amperes,

of more than 150 volts to ground

not exceeding 600 between phase conductors,

as per Section 230-95.

The wording of this section is revised for clarification, though the rule is unchanged. Two new exceptions are added where GFPE will not be required.

Exception No. 1 GFPE is not required for continuous process plants.

Exception No. 2 GFPE is not required for Fire Pumps.

Exception No. 3 GFPE on Supply side of Feeder No change.

215-11 Circuits Derived from Autotransformers

Exception No. 1 is reworded. Autotransformers are permitted to be used on new feeder circuits as well as existing feeder circuits. Previously, the rule only permitted the use of autotransformers for the extension or addition to an existing feeder.

ARTICLE 220

BRANCH-CIRCUIT, FEEDER, AND SERVICE CALCULATIONS

220-2 Computations

(b) Fraction of an Ampere *Except where computations result in a fraction of an ampere 0.5 or larger, such fractions shall be permitted to be dropped.*

Reason

This establishes a standard practice for load calculations.

Comment

This is a new statement in this article. This practice is used with the examples, which are now in Appendix D of the *Code*. Example of use when the neutral is permitted to be 70% of line current.

Neutral amps = Line amps multiplied by 70%

$N_A = L_A \times 70\%$

$N_A = 249$ amps $\times 70\% = 174.3$ amps

The 0.3 is less than 0.5 and is permitted to be dropped.

$N_A = 65$ amps $\times 70\% = 45.5$ amps

The 0.5 is not permitted to be dropped and the decimal is carried for the balance of the calculations.

Note that the new statement does not say "round up." This illustration is based on the examples in Appendix D.

220-3(b) Other Loads—All Occupancies

This section has been rearranged and given boldface titles. The reference for track lighting is deleted and some new headings are introduced. The exceptions are reworded and subdivided as part of the 1999 text. As in the 1996 *Code,* many of these statements reference other sections of the *Code* for special installations.

220-3(b)(6) Sign and Outline Lighting

Sign and outline lighting shall be computed at a **minimum** of 1200 volt-amperes for each required branch circuit.

Comment

The use of the word "minimum," indicates that the calculation is the rating of the equipment, but not less than 1200 volt-amperes.

220-3(b)(7) Show Windows

Show windows are required to be computed in accordance with either:
a. the unit load per outlet as required in other provisions of this section or
b. 200 volt-amperes per linear foot of show windows.

Reason

When the load is greater than 200 volt-amperes per linear foot, then that value should be used.

Cross Check The following is a cross check of the re-alignment of Part A. of Article 220.

Cross Check

1996	1999
Part A. General	**Part A. General**
220-1 Scope	**220-1 Scope**
220-2 Voltage	**220-2 Computations**
New	(a) Voltage
	(b) Fraction of an Ampere
220-3 Computation of Branch Circuit	**220-3 Computation of Branch Circuit**
(a) Continuous and Noncontinuous Loads	Loads
(a) Lighting Loads for Specific Occupancies	(a) Lighting Loads for Specific Occupancies
(FPN)	(FPN)
(b) Other Loads All Occupancies	(b) Other Loads All Occupancies
Exception No. 4	*Exception*
220-3(c)(1)	(1) Specific Appliance Load
220-3(c) *Exception No. 5*	(2) Electric Dryers and Household Cooking Appliances
220-3(c) *Exception No. 2*	(3) Motor Loads
220-3(c)(2)	(3) Motor Loads
220-3(c)(3)	(4) Recessed Lighting Fixtures
220-3(c)(4)	(5) Heavy Duty Lampholders
220-3(c)(6)	(6) Sign and Outline Lighting
New	(7) Show Windows
New	(a) Unit load
Part of **220-12**	(b) 200 VA per ft
220-3(c) *Exception No. 1* Part of text	(8) Fixed Multioutlet Assemblies
Exception No. 1 Part of text	(a)
Exception No. 1 Part of text	(b)
	(9) Receptacle Outlets
220-3(c)(7) Part of text	180 VA and 90 VA (new)
	(10) Dwelling Occupancies
Note from **Table 220-3(b)**	(a), (b), and (c)
220-3(c)(7) Part of text	(11) Other Outlets
220-3(d) Loads for Additions to Existing Installations	**220-3(c) Loads for Additions to Existing Installations**
(1) Dwelling Units	(1) Dwelling Units
(2) Other Than Dwelling Units	(2) Other Than Dwelling Units
210-22 Maximum Loads	**220-4 Maximum Loads**
(a) Motor-Operated and Combination Loads	(a) Motor-Operated and Combination Loads
(b) Inductive Lighting Loads	(b) Inductive Lighting Loads
(c) Range Loads	(c) Range Loads

Part B.

General

Because the title of this article includes the "Service," the phrase *"or service"* is inserted after the term "feeder" throughout part B.

220-15 Fixed Electrical Space Heating

Exception No. 2 is deleted. The exception permitted the use of an optional calculation when making standard calculations. This part of Article 220 has to do with standard calculations. Optional Calculations are covered under Part C.

Reason

The demand factors in Part C are not used when making standard calculations.

Table 220-19 Demand Loads for Household Electric Ranges

Lines have been added to this table to ease reading. Values on the table are unchanged.

CALCULATIONS—RECEPTACLES   220-3(b)(9)

SINGLE PIECE OF EQUIPMENT   *REVISED*

CALCULATIONS RECEPTACLES
SINGLE PIECE OF EQUIPMENT

90 VOLT-AMPERES PER RECPTACLE

Other Than in Dwelling Units

360 Volt-Amperes   540 Volt-Amperes

220-3(b)(9)

REVISION

The revision adds a new sentence. *A single piece of equipment consisting of a multiple receptacle comprised of four or more receptacles, shall be considered at not less than 90 volts-amperes per receptacle.*

Reason

This revision is to cover the new equipment being used with information technology equipment. When this type of equipment is installed, 90 VA is calculated for each receptacle.

Comment

The basic rule uses the term "yoke" and when two yokes were mounted in the same box, it was calculated at 360 VA. Now that the new equipment is being manufactured with a multiple of receptacles as a single piece of equipment, it needed to be clear as to how the multiple outlet equipment should be calculated.

TRACK LIGHTING

MOVED

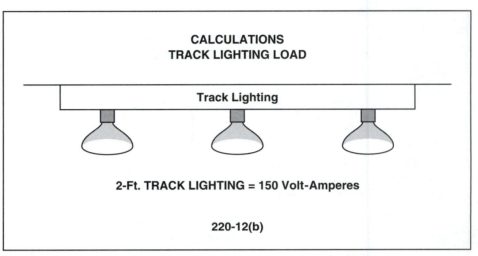

CALCULATIONS
TRACK LIGHTING LOAD

Track Lighting

2-Ft. TRACK LIGHTING = 150 Volt-Amperes

220-12(b)

MOVED FROM 410-102

Track lighting is to be calculated at 150 VA per every 2 ft. or fraction thereof. This does not apply to track lighting in dwellings, hotel or motel guest rooms.

Reason

It is appropriate for this calculations rule to be located in Article 220.

Comment

There was an *Exception* to 410-102 exempting track lighting in dwellings and hotel and motel guest rooms. This exception is made a part of the basic rule. Example of calculations for 16 ft. of track lighting.

16 ft. of track lighting ÷ 2 ft. = 8 units × 150 VA = 1200 VA.

FEEDER CALCULATIONS

SMALL APPLIANCE CIRCUIT LOADS

220-16(a) Exception

NEW

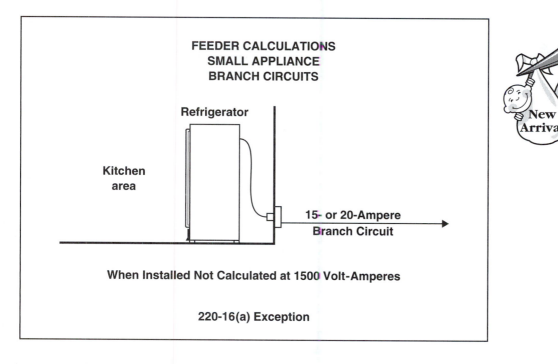

FEEDER CALCULATIONS
SMALL APPLIANCE
BRANCH CIRCUITS

Refrigerator

Kitchen
area

15- or 20-Ampere
Branch Circuit

When Installed Not Calculated at 1500 Volt-Amperes

220-16(a) Exception

BASIC RULE

The feeder calculation for the small appliance branch circuits is required to be 1500 VA for each small appliance branch circuit.

NEW EXCEPTION

When an individual small appliance branch circuit is installed for the refrigerator, it is not calculated at 1500 VA.

Reason

In the 1996 *Code*, it was not clear whether or not, when the individual branch circuit was installed for the refrigerator, this small appliance branch circuit should have a feeder calculation of 1500 VA.

Comment

This is not a required small appliance branch circuit. An exception to 210-52(b)(1) permits an individual 15- or 20-ampere small appliance branch circuit to be installed specifically for the refrigerator. This exception is listed under small appliance branch circuits; therefore, it can be identified as a small appliance branch circuit. If as a design person, you felt like 1500 VA should be used for this small appliance receptacle, the *Code* would permitted it.

FEEDER CALCULATIONS
DWELLING UNIT DRYERS
3-PHASE, 4-WIRE SYSTEM

Multifamily Dwelling 24 Units
Each Unit: One 5000 VA Electrical Dryer
Service Supply 120/208 Volts, 3-Phase, 4-Wire

24 Dryers ÷ 3 = 8 Dryers on Each Phase
8 × 2 = 16 Dryers on Adjacent Phases

Table 220-20 16 Dryers 40% Demand Factor
16 Dryers × 5 kVA × 40% = 32 kVA

32 kVA ÷ 2 = 16 kVA per Phase

Neutral
3-Phase Load 16 kVA × 3 = 48 kVA
48 kVA × 70% = 33.6 kVA

220-18

NEW RULE
Where two or more single phase dryers are supplied by a 3-phase, 4-wire feeder, the total load shall be computed on the basis of twice the maximum number connected between any two phases.

Reason
Dwelling unit electric ranges and dryers are very similar loads. This new rule puts the feeder calculation for electric dryers the same as for electric ranges.

Comment
The calculations in the illustration are based on Example D5(a) in the Appendix D.

OPTIONAL CALCULATIONS   220-30

DWELLING UNIT   *REVISED*

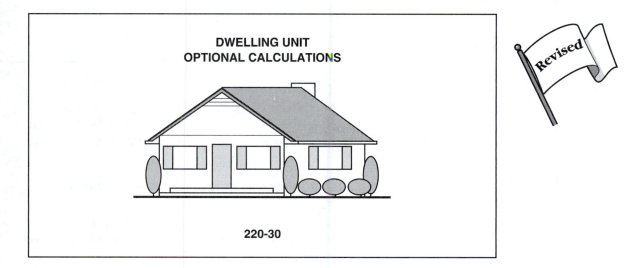

REVISION

Table 220-30 is deleted and replaced with an outlined section. The following is a brief outline of Revised Section 220-30.

220-30 Optional Calculations

(a) Feeder and Service Load (Basic requirement of where applicable not changed)

(b) General Loads 10 kVA plus 40% of the following
    (1) 1500 VA for each 2-wire appliance and laundry branch circuit
    (2) Lighting and general use receptacles 3 VA per sq. ft.
    (3) Nameplate rating of small appliances includes: range, ovens, cooking units, dryers and water heater.
    (4) *Nameplate rating of all motors and of all low-power-factor loads*

(c) Heating and Air Conditioning Loads Largest of the following six.
    (1) 100% nameplate air-conditioning and cooling
    (2) 100% nameplate heat pump compressors and supplemental heating *unless the controller prevents the compressor and heating from operating at the same time.*
    (3) 100% of thermal storage and other heating systems where the usual load is expected to be continuous.
    (4) 65% central electric space heating, including integral supplemental heating in heat pumps *unless the controller prevents the compressor and heating from operating at the same time.*
    (5) 65% of the rating with less than four separately controlled space heaters.
    (6) 40% of the rating with four or more separately controlled space heaters.

Reason

The revision is for clarification and makes the section much more user-friendly.

ARTICLE 225

225-1 Scope

(Delete) *Exception: Applicable to Electrolytic Cells.*

Reason
This is covered under special equipment.

225-6 Conductor Size and Support

There is a title change and a little rewording in the section. The exception permitting the use of messenger wire has been deleted and blended into the text. The rules for the use of messenger wire, which were in 225-13, are now part of 225-6(b). There is no change in basic intent.

225-19(d) Clearance from Buildings for Conductors Not Over 600 Volts, Nominal

(d) Final Spans. Add a new sentence: *Vertical clearances of final spans above platforms, projections, or surface from which they might be reached shall be maintained in accordance with Section 225-18.*

Reason
The revision is made to clarify the application of the rule. Section 225-18 gives the required clearances.

~~225-23 Underground Circuits~~

~~Underground circuits shall meet the requirements of Section 300-5.~~ This section is deleted as it is covered in Section 300-5.

OUTSIDE BRANCH CIRCUITS AND FEEDERS   225 PART B.

MORE THAN ONE BUILDING OR STRUCTURE   *NEW*

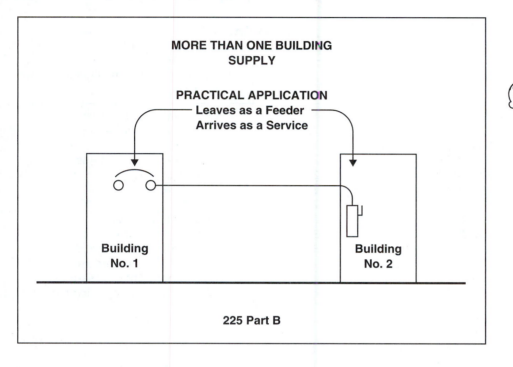

NEW PART B.

In past revisions of the *Code*, many rules for a service have been copied into Article 225 and made applicable to a second building. Now the 1999 *Code* makes a major revision and introduces a new **Part B. More Than One Building or Other Structure.** This new part duplicates all the applicable rules of Article 230 and Article 225 into one part. New section numbers are given starting with *Section 225-30. Section 225-8* of the 1996 *Code* is deleted because it was devoted to the second building or structure. It is now moved to Part B. An additional **Part C. Over 600 Volts** is also introduced. To make the move, little rewording is necessary other than to change the word "service" to "system."

Reason

The supply to a second building must now be installed under the same rules as a service, thereby giving the same safety protection to the second building as to the original building's service.

Comment

This change gives weight to the old saying, although not stated in the *Code*, "The conductors leave the first building as a feeder and arrive at the second building or structure as a service." There is very little change in the basic rule. The following is a cross check.

ARTICLE 225

New Part B. More Than One Building or Other Structure

Cross Check

1996	1999
New Part	Part B. More Than One Building or Other Structure
230-2(a)	225-30 Number of Supplies
	(a) Special Conditions
230-2(a) *Exception No. 1*	1. Fire Pump
230-2(a) Part of *Exception No. 2*	2. Emergency Systems
230-2(a) Part of *Exception No. 2*	3. Legally Required Systems
230-2(a) Part of *Exception No. 2*	4. Optional
230-2(a) Part of *Exception No. 2*	5. Parallel
230-2(a) Part of *Exception No. 3*	(b) Special Occupancy
230-2(a) Part of *Exception No. 3*	1. No Point of Access
230-2(a) Part of *Exception No. 3*	2. Sufficiently Large
230-2(a) *Exception No. 4*	(c) Capacity Requirement
230-2(a) *Exception No. 6*	(d) Different Characteristics
230-8(a) *Exception No. 1*	(e) Documented Switching Procedure
225-8(b)	225-31 Disconnecting Means
225-8(b)	225-32 Location
225-8(b) *Exception No. 1*	*Exception No. 1*
225-8(b) *Exception No. 2*	*Exception No. 2*
225-8(b) *Exception No. 3*	*Exception No. 3* (*Add towers*)
	Exception No. 4 (*New for Signs*)
230-71	225-33 Maximum Number of Disconnects
230-71(a)	(a) General
230-71(a) *Exception*	*Exception*
230-71(b)	(b) Single Pole Units
230-72	225-34 Grouping of Disconnects
230-72(a)	(a) General
230-72(a) *Exception*	*Exception*
230-72(b)	(b) Additional Disconnects
230-72(c)	225-35 Access to Occupants
230-72(c) *Exception*	*Exception*
225-8(c)	225-36 Suitable for Service Equipment
225-8(c) *Exception*	*Exception*
225-8(d)	225-37 Identification
225-8(d) *Exception No. 1*	*Exception No. 1*
225-8(d) *Exception No. 2*	*Exception No. 2*
New Title	225-38 Disconnect Construction
230-76	(a) Manual or Power Operated
230-74	(b) Simultaneous Opening of Poles
230-75	(c) Disconnect Grounded Conductor
230-77	(d) Indicating

230-79	225-39 Rating of Disconnecting Means (a) One-Circuit (b) Two-circuits (c) One-Family Dwelling (d) All Others.
225-9(b)	225-40 Access to Overcurrent Protective Devices.

New Part C. Over 600 Volts

Cross Check

1996	1999
230-203	225-50 Warning Signs
230-204(a) *Exception*	225-51 Isolating Switch *Exception*
230-205(a) 230-205(b)	225-52 Location 225-53 Type

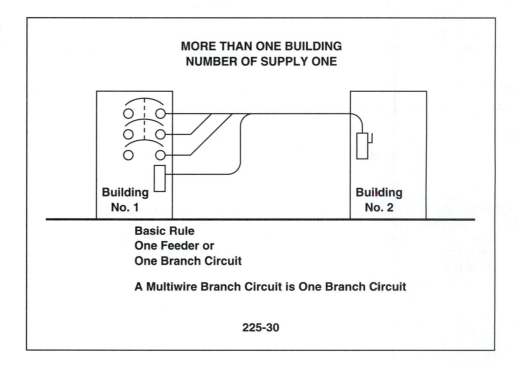

**MORE THAN ONE BUILDING
NUMBER OF SUPPLY ONE**

Building
No. 1

Building
No. 2

**Basic Rule
One Feeder or
One Branch Circuit**

A Multiwire Branch Circuit is One Branch Circuit

225-30

NEW RULE

Where a second building or structure on a property under single management is supplied by a multiwire branch circuit, it is considered as a single circuit.

Reason

The revision clarifies what the *Code* means by the term "one branch circuit." This also coordinates with 250-32 *Exception* for separate buildings or structures, permitting a single branch circuit with an equipment grounding conductor to a separate building or structure without requiring the installation of a grounding electrode at the second location.

OUTSIDE BRANCH CIRCUITS AND FEEDERS   225-31

SECOND BUILDING NUMBER OF SUPPLIES   *NEW*

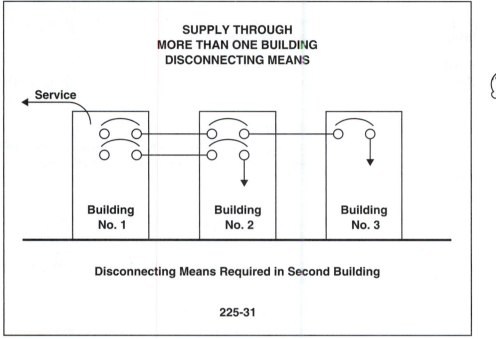

SUPPLY THROUGH
MORE THAN ONE BUILDING
DISCONNECTING MEANS

Service

Building No. 1   Building No. 2   Building No. 3

Disconnecting Means Required in Second Building

225-31

New Arrival

NEW RULE

A disconnecting means must now be provided for disconnecting all ungrounded conductors supplying, *or passing through*, a building or structure.

Reason

In case of an emergency, all electrical power in a building may have to be disconnected.

Comment

This requirement parallels the rule against the service for one building passing through another building.

ARTICLE 230

230-9 Clearance from Buildings Openings

Add a new sentence: *Vertical clearances of final spans above platforms, projections or surfaces from which they might be reached shall be maintained in accordance with Section 230-24(b).*

Reason

The revision clarifies the application of the rule. Section 230-24(b) gives the required clearances.

230-22 Insulations or Covering

Delete the following sentence: ~~Service conductors shall normally withstand exposure to atmospheric and other conditions of use without detrimental leakage of current~~

Reason

This rule is already in Section 110-11, Deteriorating Agents.

Comment

This deletion is also made for the same reason in Section 230-30 for underground Service Lateral Conductors.

230-41 Insulation of Service-Entrance Conductors

~~Service entrance conductors shall normally withstand exposure to atmospheric and other conditions of use without detrimental leakage of current.~~ The first sentence is deleted. The balance of the section remains the same.

Reason

This requirement is already covered in Section 110-11.

Delete this entire section and exception:

~~230-55 Termination of Service Equipment~~

~~Any service raceway or cable shall terminate at the inner end in a box, cabinet, or equivalent fitting that effectively encloses all energized metal parts.~~

~~*Exception: Where the service disconnecting means is mounted on a switchboard having exposed busbars on the back, a raceway shall be permitted to terminate at a busing.*~~

Reason

Chapter 3 already covers this. Several sections of Article 300 are applicable to the termination of conductors at equipment.

Delete this entire section.

~~230-63 Grounding and Bonding Service Equipment~~

This section was a cross reference of the various applicable Parts of Article 250, Grounding.

Reason

Article 250 already covers these requirements.

Delete this entire section.

230-65 Available Short Circuit Current

This section was redundant to Section 110-9, Interrupting Rating, and 110-10, Circuit Impedance and Other Characteristics.

230-71(a) Maximum Number of Disconnects

(a) General. The revision in this section now makes clear that in a single family dwelling or structure with a second set of service entrance conductors tapped off a single-service drop or lateral, six disconnecting means may be installed at the second location. The exception has been deleted and moved into the text without changing the rule.

Reason

This was not clear in the 1996 *Code*. The exception gave the intent of the *Code,* and this is now part of the basic text.

230-82 Equipment Connected on the Supply Side of Service Disconnect

Delete *Exception No. 2. Exception No. 2 was directed at meter pedestals and second buildings located away from building supply.* There were eight exceptions in the 1996 *Code.* The remaining seven exceptions are deleted and reworded, and are made a part of the text.

Reason

This deleted exception did not apply to services.

230-82 Equipment Connected to the Supply Side of Service Disconnect

This section is reworded so that only the following equipment is permitted on the supply side of the service disconnecting means. The seven exceptions are reworded and listed as new subdivisions **(1), (2), (3), (4), (5), (6), and (7)**.

230-83 Transfer Equipment and *two exceptions*

Delete this entire section.

Reason

A transfer switch suitable for service equipment would have to comply with Section 230-74, Simultaneous Opening of Poles, and this would cover transfer switch requirements.

230-91 Location of Overcurrent Protection

Delete the entire subsection.
   (b) Access to Occupants

Reason

The same rule is stated in Section 230-72(c). The rule concerns occupants having access to their source of supply. With (b) deleted, subdivision (a) is unnecessary. The information in that subdivision is now directly under the section title.

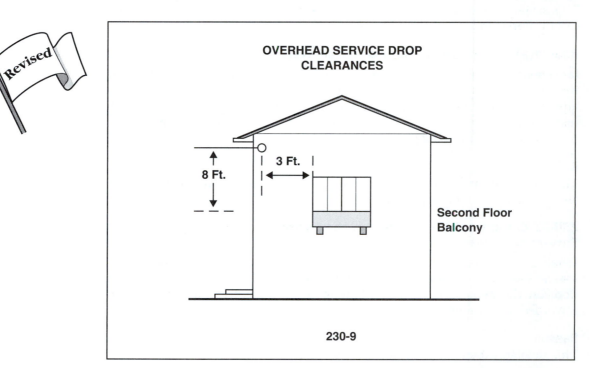

OVERHEAD SERVICE DROP
CLEARANCES

8 Ft.

3 Ft.

Second Floor
Balcony

230-9

REVISION
When the conductors are within 3 ft. *horizontally* of platforms, projections or surfaces from which they might be reached, the *clearances are required to be in accordance with Section 230-24(b).*

Reason
The revision gives a definite vertical clearance as well as a horizontal clearance requirement.

Comment
Section 230-24(b) requires 8 ft. vertical clearance with an extension of 3 ft. beyond the edge of a roof. These measurements are now applicable to balconies, platforms, or other projections.

64

SERVICE-ENTRANCE CONDUCTORS   230-40 Exception No. 4

COMMON AREA BRANCH CIRCUIT   *NEW*

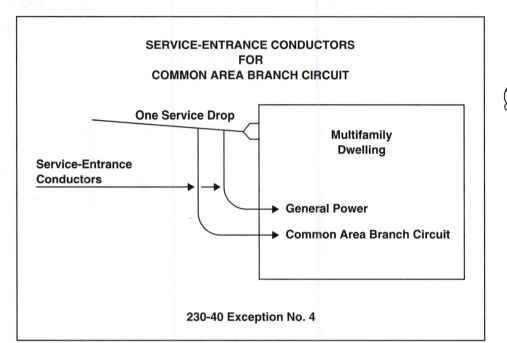

SERVICE-ENTRANCE CONDUCTORS
FOR
COMMON AREA BRANCH CIRCUIT

One Service Drop

Service-Entrance
Conductors

Multifamily
Dwelling

General Power

Common Area Branch Circuit

230-40 Exception No. 4

New Arrival

BASIC RULE

The basic rule permits one service drop or service lateral to serve one set of service-entrance conductors.

NEW EXCEPTION

A second set of service-entrance conductors is permitted for a two-family or a multifamily dwelling for the required common area branch circuit.

Reason

Separate metering of the common-area branch circuit indicates a need for this new exception.

Comment

The common area branch circuit for a multifamily or a two-family dwelling would be used for security lighting and any general alarm system that might be installed on the property.

EQUIPMENT CONNECTED TO
SUPPLY SIDE SERVICE DISCONNECT   *NEW*

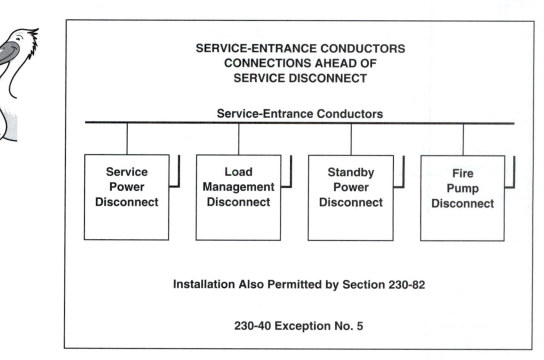

SERVICE-ENTRANCE CONDUCTORS
CONNECTIONS AHEAD OF
SERVICE DISCONNECT

Service-Entrance Conductors

| Service Power Disconnect | Load Management Disconnect | Standby Power Disconnect | Fire Pump Disconnect |

Installation Also Permitted by Section 230-82

230-40 Exception No. 5

BASIC RULE

Equipment is not permitted to be connected to the supply side of the service disconnect switch.

NEW EXCEPTION

Any or all of the following are permitted to be connected to the same service-entrance conductors ahead of the service disconnect:

Load Management Devices

Standby Power Systems

Fire Pump Equipment and Sprinkler Alarms

Reason

In the 1996 *Code,* the listed equipment was permitted to be connected ahead of the service disconnecting means in Part F of Article 230 and was not coordinated with Service-Entrance Conductors in Part D. Now the coordination is made.

Comment

230-40 Exception No. 5 refers to 230-82, which lists the different pieces of equipment that are permitted to be connected ahead of the service disconnecting means and connected to the service-entrance conductors.

MINIMUM SIZE AND RATING
UNGROUNDED CONDUCTORS   *REVISED*

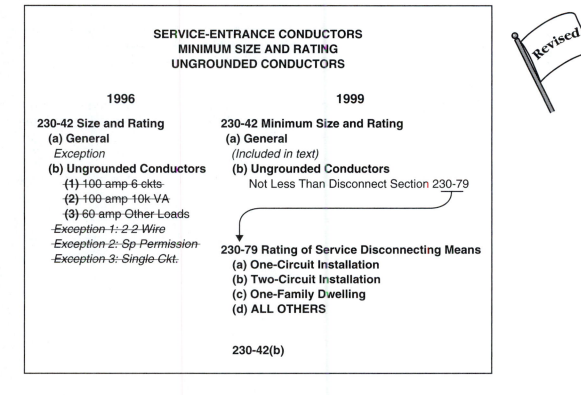

REVISION

Delete the entire section and replace with a new sentence:

   230-42(b) Ungrounded Conductors. *Ungrounded conductors are now required to have an ampacity of not less than the rating of the disconnecting means specified in Section 230-79.*

Reason

The revision reflects current building practices.

Comment

Instead of sizing the service entrance conductors based on the circuits or amperes, this revision bases them on the size of the service disconnect switch. Section 230-42 describes the sizing of the service-entrance conductors. Section 230-79 describes the sizing of the service disconnecting means. It is a general practice that when a 200-ampere service disconnecting means is used, the service-entrance conductors are rated for a minimum of 200 amperes.

   Subsections **(1)** and **(2)** were repeated in Section 230-79. They are now also deleted in Section 230-79. The following illustrates the revision of Section 230-79.

Cross Creek

1996	1999
230-79 Rating of Disconnect	**230-79 Rating of Service Disconnecting Means**

A minimum of one 100–ampere 3-wire disconnect is required. Delete the following: *(1) Where the initial installation consists of six or more 2-wire branch circuits. (2) Where the initial computed load is 10 kVA or more*.

Reason

This section discusses a specific disconnect and is therefore named in the title. The two requirements for a one-family dwelling are deleted because the demand for electricity has grown. This reflects current building practices.

SERVICE-ENTRANCE CONDUCTORS 230-46

SPLICED OR TAPPED *REVISED*

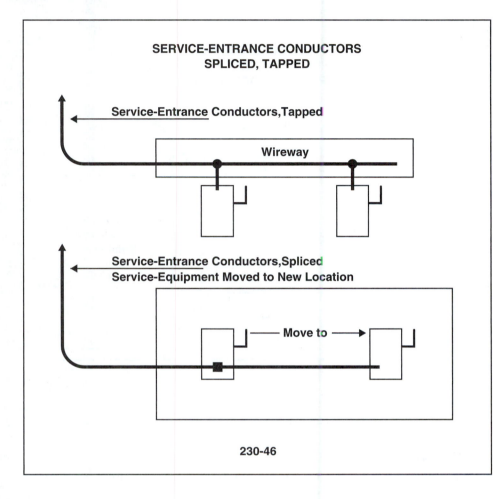

SERVICE-ENTRANCE CONDUCTORS
SPLICED, TAPPED

Service-Entrance Conductors, Tapped

Wireway

Service-Entrance Conductors, Spliced
Service-Equipment Moved to New Location

Move to

230-46

REVISION
Delete the entire Section 230-46, Unspliced Conductors, including the six exceptions and replace with new **Section 230-46, Spliced Conductors.** *Service entrance conductors are now permitted to be spliced or tapped using clamped or bolted connections. The splicing is required to comply with sections 110-14(b), 300-5(e), 300-13, and 300-15.*

Reason
This revision eliminates the six exceptions and recognizes current practices permitted by the AHJ.

Comment
Section 110-14(b) requires splices to be made with devices identified for this use and covered with insulation equal to that of the conductors being spliced. This is an example of one sentence replacing a half page of exceptions.

Section 300-5(e) has to do with underground splices.

Section 300-13 requires splices to be electrical and mechanical continuity.

Section 300-15 has to do with when a box is required.

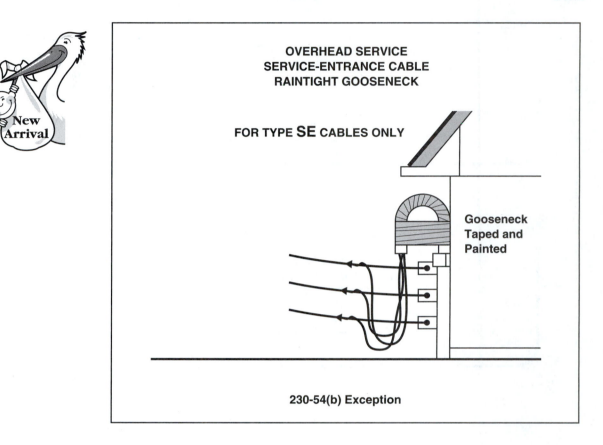

OVERHEAD SERVICE
SERVICE-ENTRANCE CABLE
RAINTIGHT GOOSENECK

FOR TYPE SE CABLES ONLY

Gooseneck
Taped and
Painted

230-54(b) Exception

BASIC RULE

Service cables shall be equipped with a raintight service head.

EXCEPTION

Type SE cable is permitted to be terminated in a gooseneck termination.

Reason

The new exception makes clear that the gooseneck installation is only recognized for use with Type SE cable.

Comment

Where the word "cables" was previously used, the literal interpretation could include Type MC or Type MI cable. It is not the intent to terminate these cables in a gooseneck termination at the service head.

SERVICE OVER 600 VOLTS NOMINAL   230-204(a)

ISOLATING SWITCH   *REVISED*

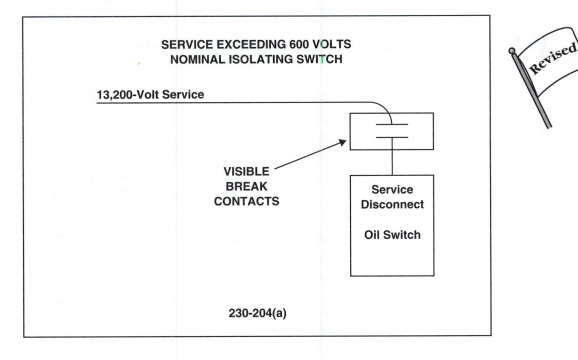

SERVICE EXCEEDING 600 VOLTS
NOMINAL ISOLATING SWITCH

13,200-Volt Service

VISIBLE
BREAK
CONTACTS

Service
Disconnect

Oil Switch

230-204(a)

REVISION

Where oil switches or air, oil, vacuum, or sulfur hexafluoride circuit breakers constitute the service disconnecting means, an air break isolating switch **with visible break contacts** shall be installed on the supply side of the disconnecting means and all associated service equipment.

Reason

This revision permits other types of isolating devices to be used rather than just an air-break isolating switch. There are SF6 or vacuum switches with visible break construction.

Comment

An isolating switch is not designed to operate under load. It is a safety switch used for isolating equipment after the circuit has been interrupted by other means. Isolating switches are used for safety and are sometimes referred to as safety switches. Therein lies the requirement for the contacts to be visible.

SERVICE OVER 600 VOLTS NOMINAL   230-205(c)

SERVICE DISCONNECT CONTROL   *NEW*

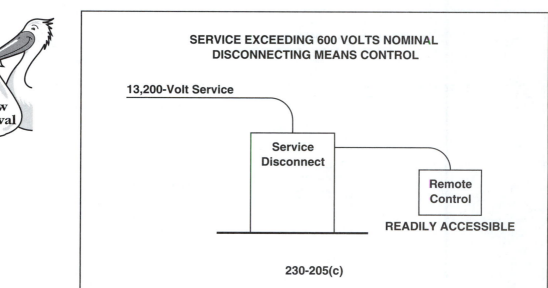

**SERVICE EXCEEDING 600 VOLTS NOMINAL
DISCONNECTING MEANS CONTROL**

13,200-Volt Service

Service
Disconnect

Remote
Control

READILY ACCESSIBLE

230-205(c)

NEW RULE

For multi-building, industrial installations under single management, the service disconnect means is permitted to be located at a separate building or structure. When so located, the service disconnect means is required to be electrically operated by a readily accessible, remote control device.

Reason

This new rule clears up the misunderstanding that a similar rule applied only to Outside Branch Circuits and Feeders, as covered in Article 225.

Comment

The 1996 *Code* deleted an exception permitting the remote control and applied it to a second building supply. This is another step in putting rules that apply to the service in Article 225, Outside Branch Circuits and Feeders, that apply to a second building.

ARTICLE 240

240-1 Scope

Insert a new part. ***Part "H" covers overcurrent protection for supervised industrial installations.***

Part "I" was Part H and still covers overcurrent protection over 600 volts nominal.

Reason

This new part to *Article 240* is added to address the concerns of industrial installations and make the *Code* more User-Friendly for industrial users.

240-3 Protection of Conductors

Considerable rearrangement has been made in *Section 240-3*, Protection of Conductors. The footnote from *Table 310-16* was moved to coordinate all overcurrent protection into *Article 240*. A definition of "Tap Conductors" is added, and Overcurrent Protection for Transformers is coordinated into one area. The following illustrates the rearrangement.

Cross Check

1996	1999
240-3 Protection of Conductors	240-3 Protection of Conductors
(a) Power Loss Hazard	(a) Power Loss Hazard
(b) Devices Rated 800 Amperes or Less	(b) Devices Rated 800 Amperes or Less
(1)	(1)
(2)	(2)
(3)	(3)
(c) Devices Rated Over 800 Amperes Footnote Table 310-16 Moved	(c) Devices Rated Over 800 Amperes
(d) Tap Conductors	(d) *Small Conductors* New
	(e) Tap Conductors *Definition* New
(i) Transformer Secondary Conductors	(f) Transformer Secondary Conductors Revised; refer to 240-21
(e) Motor-Operated Appliance Circuit Conductors	(g) Overcurrent Protection for Specific Conductor Application
(f) Motor and Motor-Control Circuit Conductors	New Table
(g) Phase Converter Supply Conductors	Alphabetical Order
(h) Air-Conditioning and Refrigeration Equipment Circuit Conductors	List Article and Section Where Overcurrent
(j) Capacitor Circuit Conductors	Protection Rules
(k) Electric Welder Circuit Conductors	Are Located.
(l) Remote-Control, Signaling, and Power Limited Circuit Conductors	
(m) Fire Alarm System Circuit Conductors	

Article 710 is deleted. The revision coordinates rules for overcurrent protection from Article 710, Over 600 Volts Nominal, into *Article 240*.

240-6 Standard Ampere Rating

(a) **Fuses and Fixed-Trip Circuit Breakers** The *Exception* following *Section 240-6(a)* was for smaller fuses. This exception is now a part of the main text.

The FPN following part (b) is now part of the main text. It indicates that the *Code* did not intend to prohibit nonstandard size fuses.

Reason
The revision eliminates as many exceptions as possible. When a FPN gives the intention of the *Code,* it is made part of the text. Whatever is the intent of the *Code* should be in the text and not in a Fine Print Note.

240-6(b) Adjustable Trip Circuit Breakers
This new subsection makes an exception part of the text. The exception permitted an adjustable trip circuit breaker, without ready access to its trip settings to be set at other than the maximum circuit breaker setting. No actual change in the rule is made.

240-8 Fuses or Circuit Breakers in Parallel
This revision deletes the exception, rewords it, and blends the rule into the text. Overcurrent devices are not connected in parallel unless in a listed factory-assembled piece of equipment. No actual change in rule.

240-12 Electrical System Coordination
This revision takes the definition for "Coordination" out of the FPN and makes it a part of the text. No actual change in rule.

240-13 Ground-Fault Protection of Equipment
The exceptions are deleted from this section and rephrased in the main text. The following illustrates the rearrangement.

Cross Check

1996	1999
240-13 Ground-Fault Protection of Equipment	240-13 Ground-Fault Protection of Equipment No change in basic requirement. New—The provisions of this section shall not apply to the disconnecting means for
Exception No. 1	(1) Industrial process
Exception No. 2	(2) Ground-fault already provided
Exception No. 3	(3) Fire pumps

Cross Check

1996	1999
240-22 Grounded Conductor No OC in grounded conductor	240-22 Grounded Conductor No OC in grounded conductor except as below
Exception No. 1	(1) No independent pole operation
Exception No. 2	(2) Motor overload protection

Comment
Change in structure but no change in intent of the rules.

240-24 Location in or on Premises
Section 240-24 has been rearranged into a more readable form. All the exceptions have been deleted, reworded, and made a part of the text without alter-

ing the intent of subsections (a), (b), (c), and (d). The change in subsection (e) will be considered separately.

Cross Check

1996	1999
240-24 Location in or on Premises	**240-24 Location in or on Premises**
(a) Readily Accessible	**(a) Accessibility**
Exception No. 1	**(1)**
Exception No. 2	**(2)**
Exception No. 3	**(3)**
Exception No. 4	**(4)**
(b) Occupant to Have Ready Access	**(b) Occupancy**
Exception No. 1	**(1)**
Exception No. 2	**(2)**
(c) Not Exposed to Physical Damage	**(c) Not Exposed to Physical Damage**
New	FPN
(d) Not in Vicinity of Easily Ignitible Material	**(d) Not in Vicinity of Easily Ignitible Material**
(e) Not Located in Bathroom	**(e) Not Located in Bathroom**

240-33 Enclosures Vertical Position

The exception is deleted, reworded, and made a part of the text. Horizontal installation of circuit breakers is permitted when the "On" and "Off" positions are clearly indicated. A new sentence is added: *Listed busway plug-in units shall be permitted to be mounted in orientations corresponding to the busway mounting position.*

Reason
The revision explicitly permits the vertical or horizontal mounting of switches on busways provided the "On" and "Off" position of the switch is clearly marked.

240-40 Disconnecting Means for Fuses

The exceptions is deleted and reworded, and is made a part of the text with no change in the intent of the rulings.

240-50 Plug Fuses General

"Shall not" is made into a positive statement "shall be permitted." The arrangement is changed but not the intent of the rule.

Section 240-60 makes the same change and rearrangement in the rules applying to cartridge fuses.

240 Part H. Supervised Industrial Installations

This is a completely new *Part H in Article 240*. The section numbers dedicated to this part range from *240-90* to *240-99*. To insert this new section as Part H, the 1996 *Code*, Overcurrent Protection Over 600 Volts, Nominal, is now changed to Part I.

Section 240-90 requires all sections of Article 240 to be fulfilled unless modified by Part H.

Reason
This revision focuses on the concerns of large industrial users of the *Code,* making it more user-friendly for them.

User Friendly Code Rule Movers
Moving From Here to There

**OVERCURRENT PROTECTION
SMALL CONDUCTORS**

Table 310-16 Footnote ——— Move to ————➤ 240-3(d)

MAXIMUM OVERCURRENT PROTECTION

No. 14 Copper Conductor	15 Amps
No. 12 Copper Conductor	20 Amps
No. 10 Copper Conductor	30 Amps
No. 12 Aluminum Conductor	20 Amps
No. 10 Aluminum Conductor	25 Amps

240-3(d)

MOVE

The footnote about overcurrent protection of the small conductors is moved from Table 310-16 to Article 240, Overcurrent Protection. The phrase *"Unless otherwise specifically permitted elsewhere in this Code,"* has been deleted. In addition to the move, there is a new Section *240-2(g), Overcurrent Protection for Specific Applications*. This new section is a table listing alphabetically when *Section 240-3(d)* is not applicable, and where overcurrent protection for specific applications can be found.

Reason

This move consolidates as many of the overcurrent protection rulings as possible under Article 240, Overcurrent Protection.

Comment

Deleting the phrase *"elsewhere in this Code"* and listing locations to which the rule does not apply eliminates controversy about applying the rule.

Section 240-2(g) is a cross reference for overcurrent protection for specific applications located elsewhere in the *Code*. It is very much like the section heading Application of Other Articles.

OVERCURRENT PROTECTION   240-3(e)

DEFINITION—TAP CONDUCTOR   *NEW*

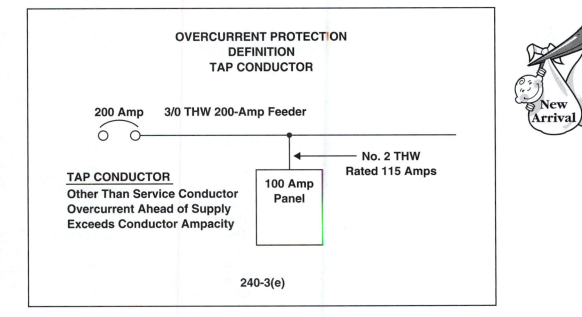

NEW DEFINITION

A tap conductor is a conductor other than a service conductor that has overcurrent protection ahead of its point of supply exceeding the value permitted for similar conductors that are protected as described elsewhere in this section.

Reason
A comparable definition was proposed for the 1996 *Code* and held for further study.

Comment
This definition does not include transformer secondary taps. They are covered in *Section 240-21*.

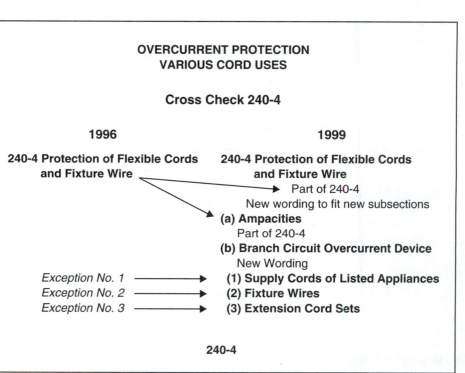

REVISION

The text is rearranged into subsections with boldface headings. The exceptions are reworded into positive statements and included in the text. The title change reflects contents of the section. The boldface heading makes it easier to locate information.

240-4(b)(3) Applies to extension cord sets and has now been extended to include *"extensions cords made with separately listed and installed attachment plugs and receptacle bodies."*

Reason

This revision makes the *Code* more User-Friendly. It also brings the field-made extension cords under the same rule as manufactured extension cords. This rearrangement makes very little change in the application of the rule.

OVERCURRENT PROTECTION 240-21(b)(1)

FEEDER TAPS *REVISED*

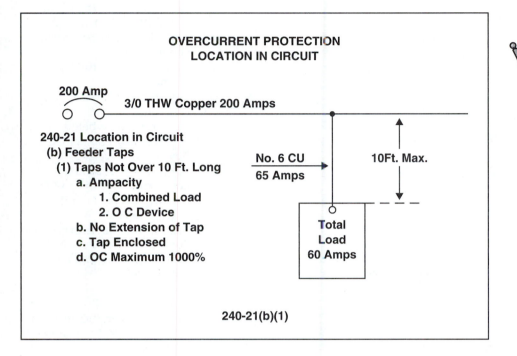

Comment

The so-called 10-ft tap rule illustrated above remains unchanged. The revision is in the layout presentation of the rule. Note the detailed outline of the various parts of the 10-ft tap rule. In the 1996 *Code,* all the information was presented in a single complex paragraph. Now all parts of the rule are readily available. Each one of the subheadings has its own easy-to-read outline of regulations. Section 240-21 has been revised and rearranged to see the various parts of each rule more clearly.

The following is a cross check between the location of the various tap rules in the 1996 *Code* and the 1999 *Code.* The complete outline for each tap rule is not outlined below, but each tap rule is outlined as illustrated above for the 10-ft tap rule.

Cross Check

1996	1999
240-21 Location in Circuit	**240-21 Location in Circuit**
(f) Branch Circuit Taps	**(a) Branch Circuit Conductors**
(a) Feeder and Branch Circuit Conductors	Part of each text
(a) Feeder and Branch Circuit Conductors	**(b) Feeder Taps**
(b) 10 Ft Tap	**(1) 10 Ft Tap**
	(a) Ampacity
	(1) Computed Load
	(2) Overcurrent Device
	(b) Termination
	(c) Field Installation 10 Times
(c) 25 Ft Tap	**(2) 25 Ft Tap**
	Subdivisions **(a), (b), and (c)**

(d) 25 Ft Primary Plus Secondary	**(3) 25 Ft Primary Plus Secondary** Subdivisions **(a), (b), (c), (d),** and **(e).**
(e) Over 25 Ft. Tap	**(4) Over 25 Ft. Tap** Subdivisions **(a), (b), (c), (d),** **(e), (f), (g), (h),** and **(i).**
(m) Outside Feeder Taps	**(5) Outside Taps of Unlimited** **Length** Subdivisions **(a), (b), (c),** and **(d).**
	(c) Transformer Secondary **Conductors** As per 1 through 4 below
240-3(i) Transformer Secondary **Conductors**	**(1) Protected by Primary OC** **Device**
240-21(b) 10 Ft Tap	**(2) Transformer Secondary** **Conductors Not Over 10 Ft** Subdivisions **(a), (1), (2), (b),** **(c),** and **FPN**
	(3) Transformer Secondary **Conductors Not Over 25 Ft** Subdivisions **(a), (b),** and **(c)**
240-21(m) Outside Feeder Taps	**(4) Outside Secondary** **Conductors** Subdivisions **(a), (b), (c),** and **(d)**
	(5) Secondary Conductors from a **Feeder Tapped Transformer**
(n) Service Conductors	**(d) Service Conductors**
(g) Busway Taps	**(e) Busway Taps**
(h) Motor Circuit Taps	**(f) Motor Circuit Taps**
(i) From Generator Terminals	**(g) Conductors From Generator** **Terminals**

BASIC RULE

Section 240-21 gives the basic rule. Overcurrent protection is required at the point where each ungrounded circuit conductor receives its supply.

REVISION

The revision is a detailed outline of **Section 240-21**, separating Branch Circuit Taps, Feeder Taps, and Transformer Taps, into separate headings and coordinating applicable rules under appropriate headings. The 10- and 25-ft feeder taps applicable to transformers are now reworded to specifically apply to transformers.

Section 240-21(c)(5) is the only new subdivision concerning transformer secondary tap protection.

Under **240-21(f) Branch-Circuit Taps** of the 1996 *Code,* the reference to Section 210-20 and Section 210-24 is deleted. As a result, the only branch-circuit tap recognized now is the one permitted for electric ranges in **210-19(b)** *Exception.*

Reason

The general rearrangement makes the *Code* easier to use. Section 210-20 and Section 210-24 are deleted because they no longer apply. The addition of Section 240-21(c)(5) recognizes overload protection for the secondary conductors, when the 25-ft rule for the primary and secondary of a transformer is used.

Comment

The person acquainted with the tap rules will find no basic change. Those unacquainted with the tap rules will find locating all the various requirements much easier.

OVERCURRENT PROTECTION   240-21(c)(5)

TRANSFORMER SECONDARY CONDUCTORS   *NEW*

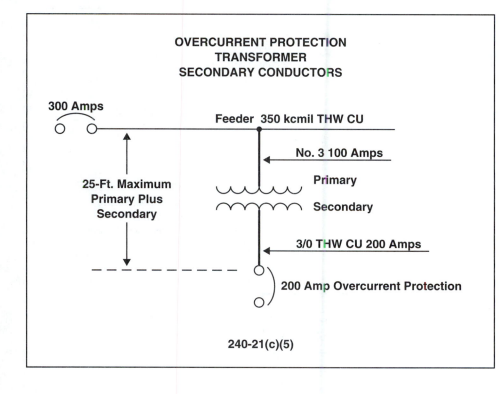

NEW FEEDER TAP RULE

Transformer secondary conductors installed in accordance with the primary plus secondary tap rule shall be permitted to terminate in overcurrent protection that will limit the current on the secondary conductors.

Reason

This new rule coordinates the required protection for the transformer secondary conductors when the permitted primary plus secondary rule is used.

Comment

The basic rule requires protection of the conductors at the point they receive their supply. This connection puts the overcurrent protection at the termination of the conductors and depends on the downstream overcurrent device to limit the current on the secondary conductors. Section 240-21(c)(5) references 240-21(b)(3). The illustration shows 240-21(b)(3) as 240-(c)(5) applies to it.

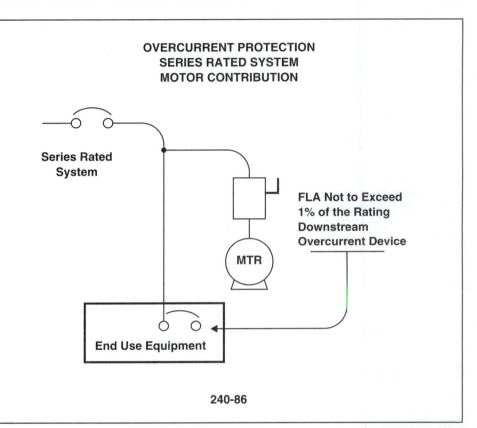

OVERCURRENT PROTECTION
SERIES RATED SYSTEM
MOTOR CONTRIBUTION

Series Rated
System

FLA Not to Exceed
1% of the Rating
Downstream
Overcurrent Device

MTR

End Use Equipment

240-86

Comment

The new rule is based on 1996 *Code,* Section 240-85, second paragraph relative to series rated circuit breakers. This new rule takes into consideration the possibility of a motor being connected between the two series rated circuit breakers.

NEW RULE

240-86. Series Rating

Where a circuit breaker does not have a fault-current rating equal to the potential fault current, and has overcurrent ahead of it in a series-rated circuit, the following rules must be observed:

(a) Marking. The combination series interrupting rating is required to be marked on the end use equipment.

(b) Motor Contribution. Series ratings are not permitted where:

(1) Motors are connected on the load side of the higher rated overcurrent device and the line side of the lower rated overcurrent device, and

(2) The sum of the motor full load currents exceed 1 percent of the interrupting rating of the lower rated circuit breaker.

Reason

In a series-rated system motors running at the time of a short circuit contribute to the fault-current at approximately the starting current (6 × FLA) of the motor.

Comment

A motor's contribution to the fault current is approximately the motor starting current or about six times the full load current. If the downstream lower rated circuit breaker has an interrupting rating of 10,000 amperes, 1 percent of 10,000 is 100 amperes. In this case, 100 amperes would be the maximum starting current of the motor.

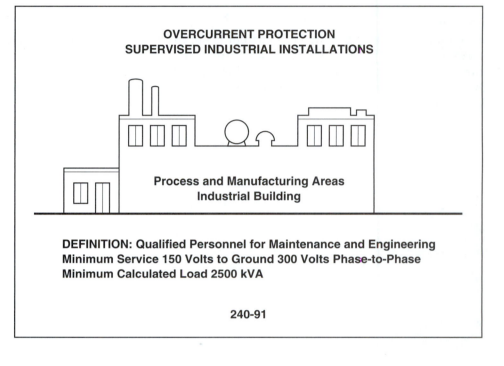

**OVERCURRENT PROTECTION
SUPERVISED INDUSTRIAL INSTALLATIONS**

**Process and Manufacturing Areas
Industrial Building**

**DEFINITION: Qualified Personnel for Maintenance and Engineering
Minimum Service 150 Volts to Ground 300 Volts Phase-to-Phase
Minimum Calculated Load 2500 kVA**

240-91

NEW DEFINITION

For the purposes of this part, "Supervised Industrial Installation" is defined as the industrial portions of a facility where all of the following conditions are met:

(1) the conditions of maintenance and engineering supervision ensure that only qualified persons will monitor and service the system;

(2) the premises wiring system has 2500 kVA or greater of load used in industrial process(es), manufacturing activities, or both, as calculated in accordance with Article 220; and

(3) the premises has at least one service that is more than 150 volts to ground and more than 300 volts phase-to-phase.

This definition would not apply to those installations in buildings used by the industrial facility for offices, warehouses, garages, machine shops, and recreational facilities that are not an integral part of the industrial plant; substation, or control center.

Comment

A good way to start out this new article is to define just what the *Code* intends for this new Part H of Article 240 to cover.

OVERCURRENT PROTECTION   240-92(c)

SUPERVISED INDUSTRIAL INSTALLATIONS
OUTSIDE FEEDER TAPS   *NEW*

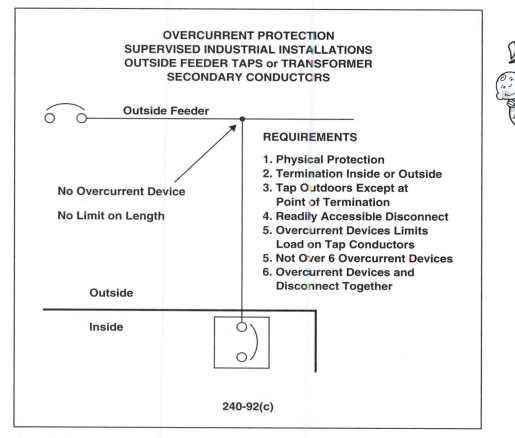

OVERCURRENT PROTECTION
SUPERVISED INDUSTRIAL INSTALLATIONS
OUTSIDE FEEDER TAPS or TRANSFORMER
SECONDARY CONDUCTORS

Outside Feeder

No Overcurrent Device

No Limit on Length

Outside

Inside

REQUIREMENTS

1. Physical Protection
2. Termination Inside or Outside
3. Tap Outdoors Except at
   Point of Termination
4. Readily Accessible Disconnect
5. Overcurrent Devices Limits
   Load on Tap Conductors
5. Not Over 6 Overcurrent Devices
6. Overcurrent Devices and
   Disconnect Together

240-92(c)

NEW RULE

Outdoor feeder taps or transformer secondary taps are permitted without overcurrent protection at the point the conductor receives its supply when all the following conditions are met:

(1) Conductors are suitably protected.

(2) Terminate in not more than six overcurrent devices grouped together. The sum of the overcurrent devices will limit the current on the tap.

(3) Installed outdoors except at point of termination.

(4) Overcurrent device is integral part of disconnect or located immediately adjacent to it.

(5) The disconnecting means is permitted to be installed inside or outside the building at a readily accessible location. When installed inside, it is required to be installed nearest the entrance of the conductors to the building.

Reason

This new rule establishes a workable way to manage special industrial installations and lists the rules separately from the main body of the *Code*.

Comment

The new rule is primarily directed at large transformers installed on an industrial premises. For large installations, when the outside conductors are terminated in indoor switchgear, it is extremely difficult to terminate the conductor when limited to 25 ft.

OVER 600 VOLTS NOMINAL
FEEDERS AND BRANCH CIRCUITS *MOVED*

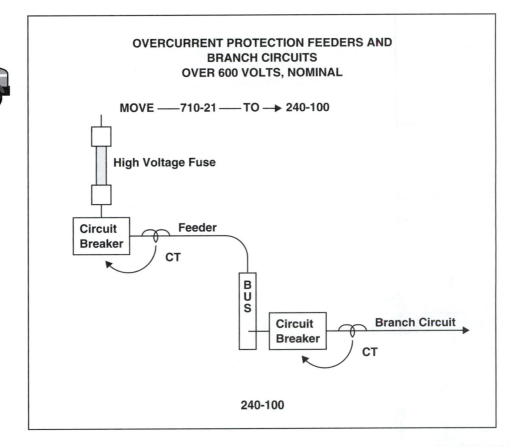

OVERCURRENT PROTECTION FEEDERS AND
BRANCH CIRCUITS
OVER 600 VOLTS, NOMINAL

MOVE ——710-21—— TO —➤ 240-100

High Voltage Fuse

Circuit Breaker

Feeder

CT

BUS

Circuit Breaker

Branch Circuit

CT

240-100

Comment

Article 710, Over 600 Volts, Nominal, has been deleted from the 1999 *Code*. The rules previously in Article 710 have been moved to other applicable articles, and a new Article 490, Over 600 Volts, Nominal, has been added. Most of the rules in Article 710 are now in Article 490. The rule concerning overcurrent protection has been moved to Section 240-100.

Part H, Over 600 Volts Nominal, of the 1996 *Code* has been deleted. An entirely revised Part I, Over 600 Volts Nominal, is introduced with rules from Article 240 Part H and former Section 710-20 blended together without alteration of intent. The previous exceptions are now part of the text.

RULE MOVED

240-100 Feeder and Branch Circuits
(a) Overcurrent protection is required in each ungrounded feeder or branch-circuit conductor.
  (1) The overcurrent device is permitted to be a circuit breaker actuated through the operation of a current transformer and relay.
  (2) Fuses are also permitted for the overcurrent protection.
(b) The overcurrent device must be capable of interrupting all values of current that may be at their point.
(c) The overcurrent device is also required to protect the conductors from damage.

Reason

This change is made by the task group and the Correlating Committee to make the *Code* more user friendly.

Comment

The revisions make the rules common to feeders and branch circuits, listing different ways that current transformers and relays can be used to trip the circuit breaker when they are used for the overcurrent protection. The installation and other regulations that apply to circuit breakers and fuses installed on over 600 volts, nominal, circuits are covered in the new Article 490.

ARTICLE 240 OVERCURRENT PROTECTION PART I.

Cross Check

1996	1999
Part H. Overcurrent Protection Over 600 Volts, Nominal	Part I. Overcurrent Protection Over 600 Volts, Nominal

240-100 Feeders
240-101 Branch Circuits
  Each ungrounded conductor
	240-100 Feeders and Branch Circuits
	
	  (a) Each ungrounded conductor or one of the following

710-20 Overcurrent Protection
  (a) Overcurrent Relays and Transformers
    Exception No. 1
    Exception No. 2
  (b) Fuses
	
	    (1) Overcurrent Relays and Transformers
	      Included in text
	      Included in text
	    (2) Fuses

240-100 Feeders
240-101 Branch Circuits
240-100 (FPN)
 	
240-100 Feeders
 	
  Exception (Fire Pump)
240-100 Feeders
	(b) Protective Devices
	
	(c) Conductor Protection
	240-101 Additional Requirements for Feeders
	  (a) Rating or Setting of the Overcurrent Device
	    Included in text
	  (b) Feeder Taps

ARTICLE 250

Goal

The rearrangement is one of the biggest changes in the 1996 *Code* as extensive moving around of rules has taken place. The goal in revising and restructuring Article 250 is to make it more user friendly by coordinating applicable rules into one location. In so doing, many rules have been moved, and the only section number to remain the same in the 1999 *Code* as in the 1996 *Code* is Section 250-1, Scope. After passing Section 250-1, every section in Article 250 has a new section number. The number of actual *Code* rule changes in Article 250 is minimal.

Rearrangement	Section Numbering
Part A. General	250-1 through 250-12
Part B. Circuit and System Grounding	250-20 through 250-36
Part C. Grounding Electrode System and Grounding Electrode Conductor	250-50 through 250-70
Part D. Enclosure, Raceway, and Service Cable Grounding	250-80 through 250-86
Part E. Bonding	250-90 through 250-106
Part F. Equipment Grounding and Equipment Grounding Conductor	250-110 through 250-126
Part G. Methods of Equipment Grounding	250-130 through 250-148
Part H. Direct Current	250-160 through 250-169
Part J. Meters and Relays	250-170 through 250-178
Part K. Grounding of Systems and Circuits 1 kV and Over (High Voltage)	250-180 through 250-190

Things to Watch For

One of the keys to following the revisions in Article 250 is to watch the way the sections have been outlined to make it easier to find and apply the rule back to the section heading under which it is given.

Rules with a general application to all bonding, grounding conductor connections, and objectionable current rules have been moved up into Part A. General.

Two new terms have been introduced:

~~Effective Ground Path~~ is replaced with **Performance of Fault Current Path**.

~~Lighting Rod Conductors~~ is replaced with **Air Terminals**.

Several changes take place in the handling of 1996 exceptions without altering the intent of the rules. While the goal of eliminating as many exceptions as possible was accomplished, many other exceptions remain just as they were in the 1996 *Code,* but they are relocated as follows:

1. Many exceptions have been reworded and included in the main text.
2. Where there were a number of exceptions, the exceptions may be reworded and given a subdivision heading for ease of location and identification.
3. A few exceptions that addressed obsolete material are deleted.
4. A few exceptions that were in conflict with other parts of the *Code* or duplicated a rule expressed elsewhere in the *Code* are deleted

Long sections are now outlined with easy-to-locate headings. Some rules that combined a couple of items in the same rule are separated into two sub-headings for clarity.

Part J. of the 1996 *Code* was entitled "Grounding Conductors" and included the "Grounding Electrode Conductor" and the "Equipment Grounding Conductor." Part J. has been deleted and the rules distributed between **Part C. Grounding Electrode Systems and Grounding Electrode Conductor** and **Part F. Equipment Grounding and Equipment Grounding Conductor.**

A real effort has been made to reduce to a minimum the use of the term "Grounding Conductor" and replace it with the term "Grounding Electrode Conductor" or "Equipment Grounding Conductor." The term "System Grounding Conductor" identifies the conductor at a separate building or structure installed between the grounding electrode at the separate building and the disconnecting means for the building supply.

Part K. Grounding Conductor Connections of the 1996 *Code* has been deleted and the rules redistributed according to the rule's application. If they have a general application, they moved to Part A. General. If they have a specific application, they moved to the specific applicable location.

PERFORMANCE FAULT CURRENT PATH *REVISED*

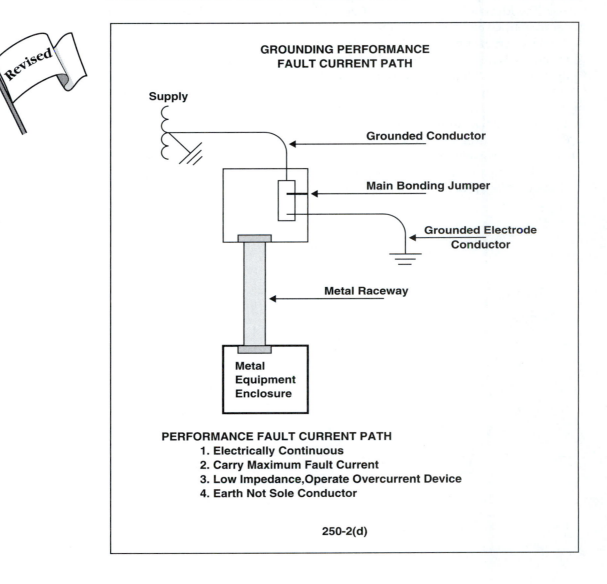

GROUNDING PERFORMANCE
FAULT CURRENT PATH

Supply

Grounded Conductor

Main Bonding Jumper

Grounded Electrode
Conductor

Metal Raceway

Metal
Equipment
Enclosure

PERFORMANCE FAULT CURRENT PATH
1. Electrically Continuous
2. Carry Maximum Fault Current
3. Low Impedance,Operate Overcurrent Device
4. Earth Not Sole Conductor

250-2(d)

REVISION

The revision to Section 250-2 consists of introducing a new heading and a new term and making the FPNs following Section 250-1 into part of the text of this new section.

> *250-2 General Requirements for Grounding and Bonding.*
>
> *(a) Grounding of Electrical Systems*
> *(b) Grounding of Electrical Equipment*
> *(c) Bonding of Electrical Conductive Material and Equipment*
> *(d) Performance of Fault-Current Path*

The new term introduced here *is "Performance of Fault-Current Path."*

Reason

This revision puts the FPN into the text making it an enforceable part of the *Code*. The new term, "Performance of Fault Current Path," is a rewording and updating of the section called "Effective Grounding" in the 1996 *Code*.

GROUNDING   250-6(e)

OBJECTIONABLE DIRECT-CURRENT (DC) CURRENTS   *NEW*

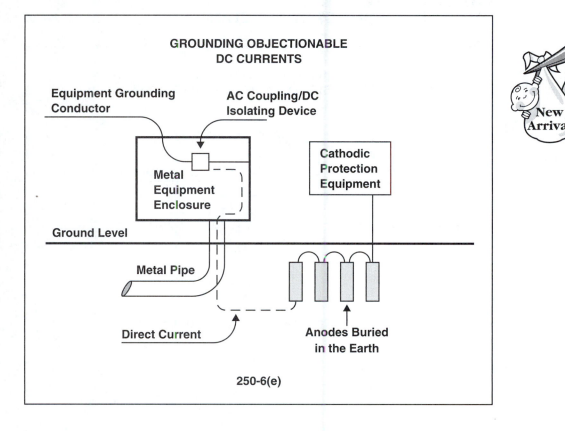

NEW RULE

250-6(e) Isolation of Objectionable DC Ground Currents. Where isolation of objectionable dc ground currents from cathodic protection systems is required, a listed ac coupling/dc isolating device is permitted in the equipment grounding path to provide an effective return path for ac ground-fault currents while blocking dc current.

Reason

There is the possibility of dc currents entering through grounded metal equipment. The ac coupling/dc isolating device protects against this possibility.

Comment

The ac coupling/dc isolating device will block the flow of dc current back into the system while letting the ac current flow in the equipment grounding conductors. This piece of equipment is tested and listed for this purpose.

GROUNDING ELECTRODE
CONDUCTOR POINT OF CONNECTION

NEW

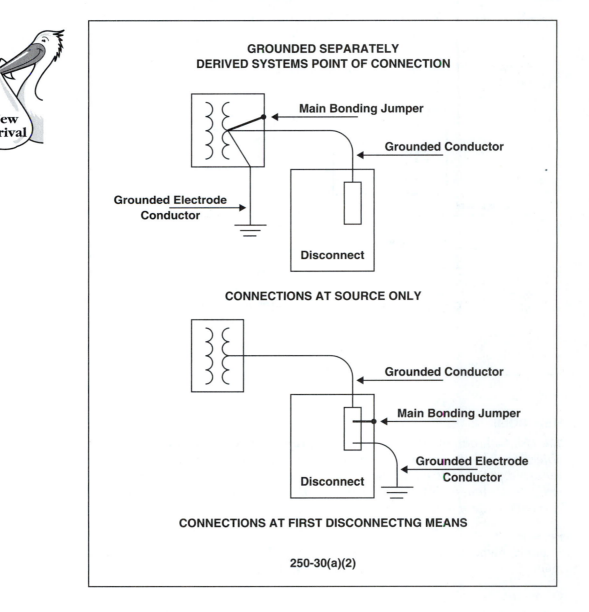

GROUNDED SEPARATELY
DERIVED SYSTEMS POINT OF CONNECTION

Main Bonding Jumper

Grounded Conductor

Grounded Electrode Conductor

Disconnect

CONNECTIONS AT SOURCE ONLY

Grounded Conductor

Main Bonding Jumper

Grounded Electrode Conductor

Disconnect

CONNECTIONS AT FIRST DISCONNECTNG MEANS

250-30(a)(2)

NEW RULE

The grounding electrode conductor is required to be connected at the same point that the bonding jumper is installed.

Reason

This new rule establishes a one-point grounding system for the installation of the bonding jumper and the grounding electrode conductors for a separately derived system.

Comment

This same requirement is repeated for coordination under Bonding Jumper in Section 250-30(a)(1).

The previous rule permitted the grounding electrode to be connected at the source or at the first disconnecting means and the main bonding jumper at the first disconnecting means. This resulted in two different connections. A grounding electrode conductor connected at the source and the main bonding jumper installed at the first disconnecting means using ground-fault protection could give a problem. The size of the grounding electrode conductor is still based on the size of the phase conductors.

GROUNDING—
SEPARATELY DERIVED SYSTEM

250-30(a)(3)(b)

GROUNDING ELECTRODE

REVISED

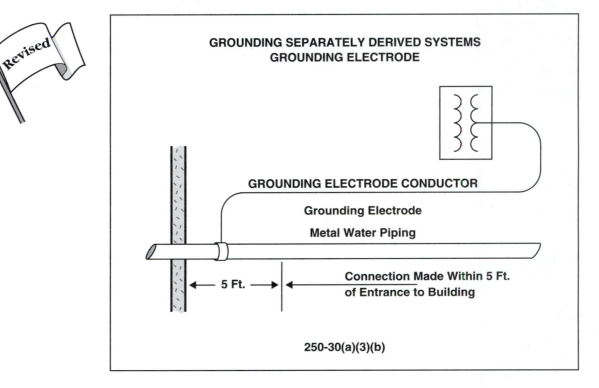

Revised

GROUNDING SEPARATELY DERIVED SYSTEMS
GROUNDING ELECTRODE

GROUNDING ELECTRODE CONDUCTOR

Grounding Electrode

Metal Water Piping

5 Ft.

Connection Made Within 5 Ft.
of Entrance to Building

250-30(a)(3)(b)

REVISION OF BASIC RULE

The grounding electrode for a separately derived system is required to be the nearest one of the following:

(a) effectively grounded building steel,

(b) effectively grounded metal water pipe within 5 feet from the point of entrance to the building, or

(c) other existing or made electrodes.

NEW EXCEPTION

250-30(3)(b). Exception: In industrial and commercial buildings with proper maintenance and supervision, the grounding electrode for a separately derived system is permitted to be an effectively grounded metal water pipe without regard to the water pipe's distance from its point of entry to the building, where the portion of the water pipe used for the electrode is exposed.

Reason

This revision and new exception clarify the use of an effectively grounded metal water pipe as the grounding electrode.

Comment

The first choice here for a grounding electrode is the effectively grounded building steel. The second choice is an effectively grounded metal water pipe with limitations. The following conditions apply:

1. The grounding electrode conductor is installed back to and connected within 5 ft of where the effectively grounded metal water pipe enters the building.
2. When all the conditions listed are met for an industrial or commercial building, the grounding electrode conductor is permitted to be connected to the effectively grounded water pipe anywhere along the exposed metal water line.

The third choice is some other grounding electrode, as permitted in Article 250.

GROUNDING—
SEPARATELY DERIVED SYSTEM   250-30(a)(3)(c) Exception

GROUNDING ELECTRODE   *NEW*

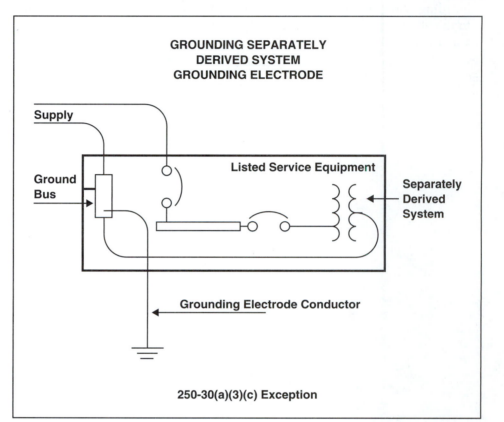

BASIC RULE

The grounding electrode is required to be effectively grounded building steel, effectively grounded water pipe, or other made electrode.

NEW EXCEPTION

Exception for (a), (b), and (c): Where a separately derived system originates in listed equipment also used as service equipment, the grounding electrode used for the service equipment shall be permitted as the grounding electrode for the separately derived system, provided the grounding electrode conductor from the service equipment to the grounding electrode is of sufficient size for the separately derived system. Where the equipment ground bus internal to the service equipment is not smaller than the required grounding electrode conductor, the grounding electrode connection for the separately derived system shall be permitted to be made to the bus.

Reason

This new exception covers present installations considered to be safe.

GROUNDING—UNGROUNDED   250-30(b)

SEPARATELY DERIVED SYSTEM   *NEW*

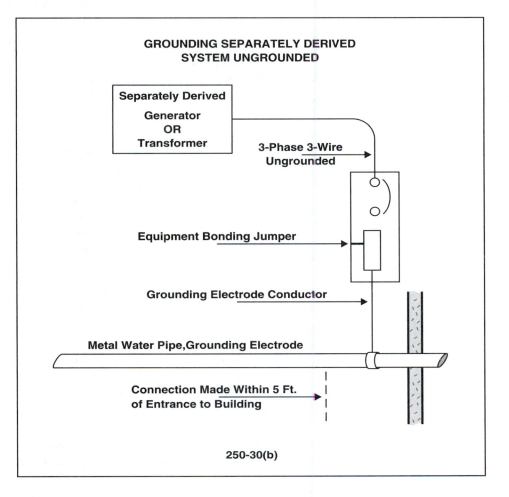

GROUNDING SEPARATELY DERIVED
SYSTEM UNGROUNDED

Separately Derived
Generator
OR
Transformer

3-Phase 3-Wire
Ungrounded

Equipment Bonding Jumper

Grounding Electrode Conductor

Metal Water Pipe,Grounding Electrode

Connection Made Within 5 Ft.
of Entrance to Building

250-30(b)

NEW RULE

An ungrounded separately derived system is required to meet the following requirements:

1. *Grounding electrode conductor sized as per Table 250-66.*
2. *The enclosure is required to be bonded to the grounding electrode conductor.*
3. *The connection is permitted to be made at any point from the source to the first disconnecting means.*
4. *The grounding electrode needs to be as near as possible and is permitted to be:*
   1. *The nearest effectively grounded building steel.*
   2. *An effectively grounded water pipe with the connection made within 5 ft of where the water pipe electrode enters the building.*
   3. *A made electrode is permitted when neither of the above two is available.*
5. *Grounding methods are the same as required for any other grounding electrode and grounding electrode conductor installation.*

Reason

This particular installation was not clearly covered in the previous edition of the *Code*. The requirements parallel the requirements for grounding a separately derived ac system.

GROUNDING SECOND BUILDING
GROUNDED SYSTEM
GROUNDING-ELECTRODE CONDUCTOR
CONNECTION

Service

With Phase Conductors

Building No. 1

Equipment Grounding Conductor

Grounded Conductor

Building No. 2

NO Connection Between Buses

250-32(b)(1)

NEW RULE

When an equipment grounding conductor is installed with the circuit conductors for grounding equipment in the separate building or structure, the grounded conductor is not permitted to be connected to the grounding electrode at the separate building.

Reason

The new rule clarifies the permitted connections at the second building or structure.

Comment

When the equipment grounding conductor and the grounded conductor are run to the second building and connected to the grounding electrode in the second building, the neutral conductor loses control over normal electron flow of the circuit. Instead of all the normal circuit neutral current flowing back on the neutral circuit conductor, the electrons can flow on a multitude of paths.

GROUNDING—SECOND BUILDING   250-32(b)(2)

GROUNDING CONNECTIONS   *NEW*

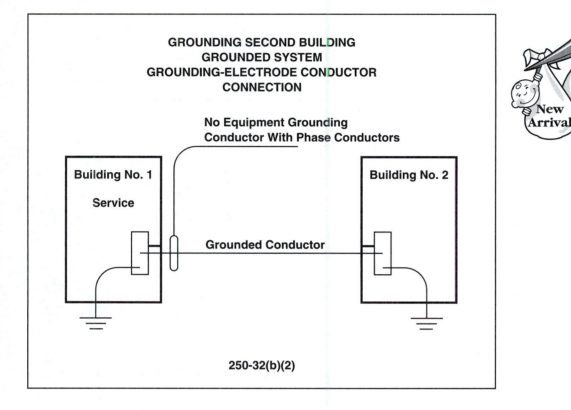

**GROUNDING SECOND BUILDING
GROUNDED SYSTEM
GROUNDING-ELECTRODE CONDUCTOR
CONNECTION**

No Equipment Grounding
Conductor With Phase Conductors

Building No. 1

Service

Grounded Conductor

Building No. 2

250-32(b)(2)

NEW RULE

When no equipment grounding conductor is installed to a separate building or structure, the grounded conductor is required to be connected to the grounding electrode system provided:

1. *There is no continuous metallic path bonded to the grounding systems in both buildings involved, and/or*
2. *There is no ground-fault protection of equipment installed on the common service*

Reason

The new rule protects the operation of the service ground-fault protection on the common service. The previous *Code* did not cover installation.

Comment

Where there is a continuous metal path between the two buildings or structures, or ground-fault protection on the common service, the grounded conductor would not be connected to the grounding electrode system. The installation then would be the same as in the previous illustration.

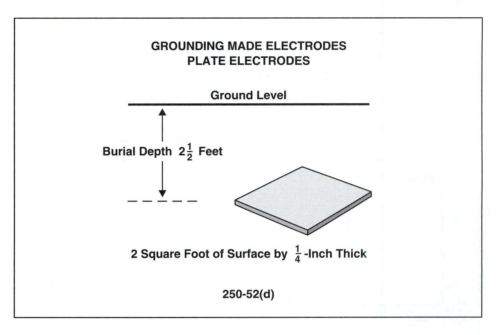

GROUNDING MADE ELECTRODES
PLATE ELECTRODES

Ground Level

Burial Depth $2\frac{1}{2}$ Feet

2 Square Foot of Surface by $\frac{1}{4}$-Inch Thick

250-52(d)

NEW RULE

A plate electrode is required to be not less than $2^1/_2$ ft below the surface of the earth.

Reason
The statement, "Where practical, made electrodes are embedded below the permanent moisture level," was too vague for enforcement. Now a specific depth is established.

Comment
This new requirement adds to the existing rule giving the size of the plate electrode.

METAL WATER PIPING
SEPARATELY DERIVED SYSTEM   *REVISED*

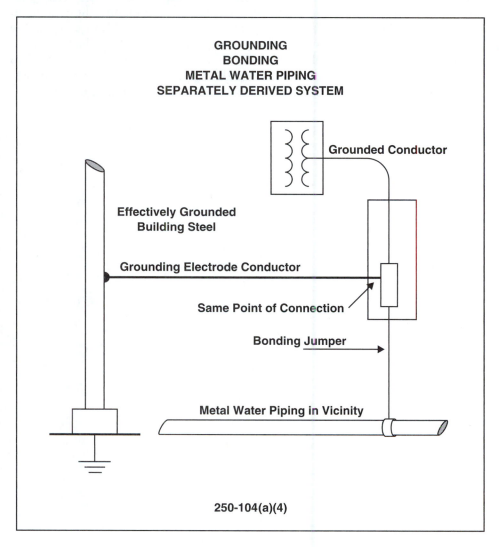

GROUNDING
BONDING
METAL WATER PIPING
SEPARATELY DERIVED SYSTEM

Grounded Conductor

Effectively Grounded
Building Steel

Grounding Electrode Conductor

Same Point of Connection

Bonding Jumper

Metal Water Piping in Vicinity

250-104(a)(4)

REVISION

The grounded conductor of a separately derived system is required to be bonded to the nearest available point on the interior metal water piping. The bonding connection is required to be made at the same point as the grounding electrode conductor is connected. The bonding jumper is sized according to Table 250-66.

Reason

By requiring the bonding to be at the nearest available point and giving specific instructions for sizing and installation, this revision should clarify the confusion caused by the wording in the 1996 *Code*.

Comment

The illustration shows the grounding electrode conductor connected to the grounded conductor at the panel. If the grounding electrode conductor had been connected to the grounded conductor at the transformer, the bonding jumper would also have to be connected at the transformer.

This rule is located under bonding and is treated as a bonding jumper. However, Table 250-66 is used for sizing both this bonding jumper and the grounding electrode conductors, indicating that this bonding jumper is sized just as if it were a grounding electrode conductor. Thus, if the effectively grounded building steel is used as the grounding electrode and there is metal water pipe in the vicinity, a bonding jumper is required. Alternately, if the effectively grounded metal water pipe is used as the grounding electrode conductor and the grounding electrode conductor is installed and connected to the first 5 ft of metal water pipe entering the building, the bonding jumper is still required in case the metal continuity of the water pipe is interrupted.

LIGHTNING PROTECTION SYSTEM *DELETED—REVISED*

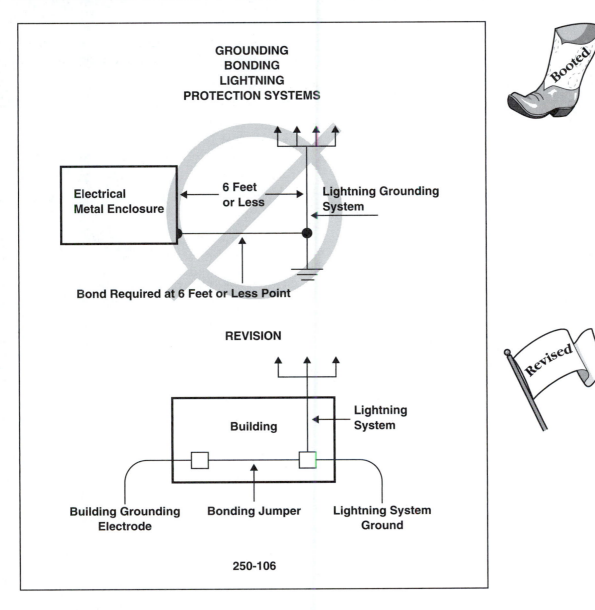

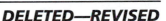

DELETION

~~250-46 Spacing from Lightning Rods.~~ This entire section requiring bonding of equipment and enclosures to be kept at least 6 ft away from lightning rod conductors or to be bonded to the lightning rod conductors at that point is deleted.

REVISION

This section is replaced with *Section 250-106, Lightning Protection System. The lightning protection system ground terminals are now required to be bonded to the building or structure grounding electrode system.*

NEW FINE PRINT NOTE

The new (FPN No. 2) refers to NFPA 780, which contains regulations for lighting protection equipment spacing.

Reason

This protection of equipment can be better served by utilizing the safety measurements in NFPA 780 Standard for the Installation of Lightning Protection Systems.

EQUIPMENT FASTENED IN PLACE   *EXCEPTION—DELETION*

BASIC RULE

Equipment operating with any terminal over 150 volts to ground is required to be grounded. The *Code* section number for this deleted exception in the 1996 *Code* is 250 42(f) *Exception No. 1.*

EXCEPTION—DELETED

250-42(f) *Exception No. 1: Enclosures for switches or circuit breakers used for other than service equipment and accessible to qualified person only.*

Reason

Failure to ground metal electrical enclosures over 150 volts to ground can be dangerous. Electricity cannot tell the difference between a qualified person and an unqualified person. Electricity has killed both.

Result of Deletion

Now all equipment under or over 150 volts to ground is required to be grounded.

ELECTRICAL EQUIPMENT ON SKIDS *NEW*

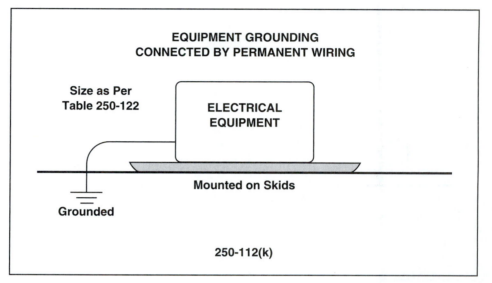

**EQUIPMENT GROUNDING
CONNECTED BY PERMANENT WIRING**

Size as Per
Table 250-122

**ELECTRICAL
EQUIPMENT**

Mounted on Skids

Grounded

250-112(k)

NEW RULE

Permanently mounted electrical equipment and skids are required to be grounded with an equipment-bonding jumper sized as per Section 250-122.

Reason

This installation was not previously covered in the *Code.*

Comment

For example, a large piece of electrical equipment is often factory-mounted on skids for shipping and moving. When it arrives on the job site, it is permanently installed with the skids.

EQUIPMENT GROUNDING CONDUCTOR   250-122(f)(2)

PARALLEL CONDUCTOR—SIZE   *REVISED*

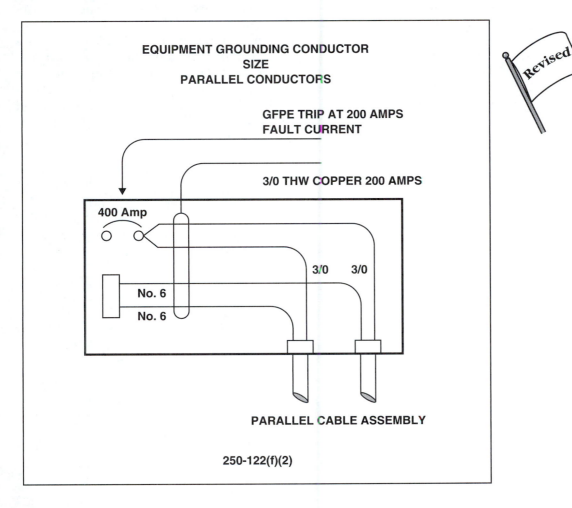

EQUIPMENT GROUNDING CONDUCTOR
SIZE
PARALLEL CONDUCTORS

GFPE TRIP AT 200 AMPS
FAULT CURRENT

3/0 THW COPPER 200 AMPS

400 Amp

3/0   3/0

No. 6

No. 6

PARALLEL CABLE ASSEMBLY

250-122(f)(2)

GENERAL RULE

Where conductors are run in multiple raceways or cables, the size of each equipment grounding conductor, in each raceway or cable, is required to be based on the ampere rating of the overcurrent-protective device.

REVISION—Specific Application

Where the conditions of maintenance and supervision ensure that only qualified persons will service the installation, the following sizing of the equipment grounding conductor is permitted for the installation of parallel cables:

1. The ground-fault protection equipment is required to be set to trip the overcurrent protective device at not more than the ampacity of a single ungrounded phase conductor in a cable

2. The ground-fault protective device must be listed for the purpose.

Reason

Cable assemblies with oversize equipment grounding conductors are not manufactured.

Comment
In the illustration, using the general rule for parallel conductors, the equipment grounding conductor in each raceway or cable would be based on the 400-ampere overcurrent protection, and there would be a No. 3 copper equipment grounding conductor in each raceway.

For this special installation, the rated ampacity of 3/0 THW copper in a cable is 200 amps. Based on the manufacturer's assumption that the 3/0 conductors will have overcurrent protection at not over their 200-ampere rating, a No. 6 copper equipment grounding conductor is manufactured in the cable. Its use is now permitted where parallel cable assemblies are installed, provided it is listed for the purpose and GFPE is also installed.

GROUNDING DIRECT-CURRENT   250-169

SEPARATELY DERIVED UNGROUNDED SYSTEM   *NEW*

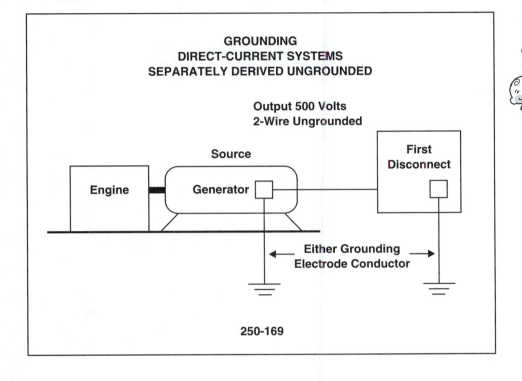

NEW RULE

A separately derived ungrounded direct current system (such as an engine generator set) is required to have a grounding electrode and a grounding electrode conductor connected at the source or first disconnecting means to provide for grounding metal enclosures. The conductor is sized according to Section 250-166.

Reason
This rule is necessary because this type of installation was not covered in the previous *Code.*

EQUIPMENT GROUNDING CONDUCTORS SEPARATELY DERIVED SYSTEM

REVISED

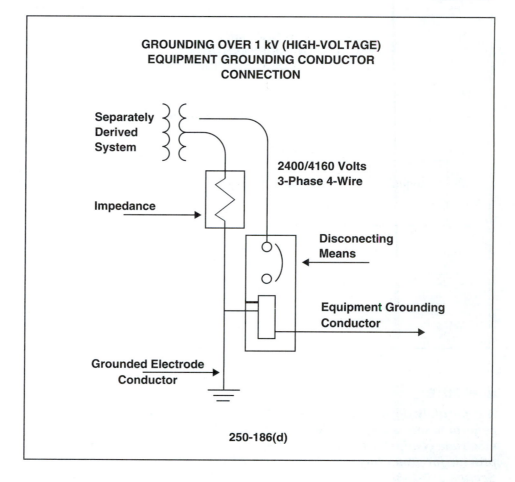

GROUNDING OVER 1 kV (HIGH-VOLTAGE) EQUIPMENT GROUNDING CONDUCTOR CONNECTION

250-186(d)

REVISION

The equipment grounding conductor for an impedance grounded system operating at over 1 kV is required to have the equipment grounding conductor connected to the grounding electrode conductor at the disconnecting means for a separately derived system.

Reason

This revision results from the increased use of the high voltage separately derived system on wiring systems of large premises.

Comment

The equipment grounding conductor is connected to the grounding electrode conductor on the downstream side of the impedance.

CHAPTER 3
WIRING METHODS AND MATERIALS

■ ■

ARTICLE 300

WIRING METHODS

300-1 Scope

The seven exceptions are deleted:

Exception No. 1	Referenced Article 504	Considered covered in 90-3
Exception No. 2	Referenced Article 725	Relocated to Article 725
Exception No. 3	Referenced Article 760	Relocated to Article 760
Exception No. 4	Referenced Article 770	Relocated to Article 770
Exception No. 5	Referenced Article 800	Considered covered in 90-3
Exception No. 6	Referenced Article 810	Considered covered in 90-3
Exception No. 7	Referenced Article 820	Considered covered in 90-3

Reason

The exceptions were redundant because the information is presented elsewhere in the *Code*.

300-18 Raceway Installations

New rule added: **(b) Welding.** *Metal raceways are not permitted to be terminated or supported by welding unless specifically permitted elsewhere in this Code.*

Reason

The heat from welding destroys the finish inside and outside the metal raceway.

OVER 600 VOLTS, NOMINAL WITH
OTHER VOLTAGE CONDUCTORS   *MOVED AND REVISED*

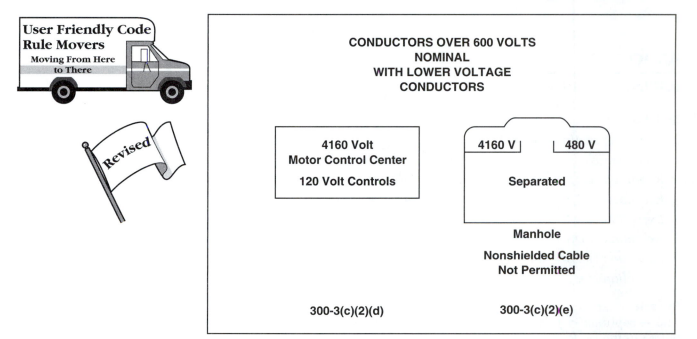

**User Friendly Code
Rule Movers**
Moving From Here
to There

Revised

CONDUCTORS OVER 600 VOLTS
NOMINAL
WITH LOWER VOLTAGE
CONDUCTORS

| 4160 Volt
Motor Control Center

120 Volt Controls |

| 4160 V | 480 V |

Separated

Manhole

**Nonshielded Cable
Not Permitted**

300-3(c)(2)(d)   300-3(c)(2)(e)

MOVE

Following 300-3 delete the FPN that referenced Section 300-32.

Section 300-32 is deleted and replaced with a statement referencing Section 300-3. The two exceptions from Section 300-32 are moved and reworded, and they are now part of Section 300-3 text.

REVISION

The exceptions to Sections 300-3 and 300-32 are deleted and made a part of the text. The text is reworded, rearranging it into an outline without changing the intent of the existing rules. One new statement is added:

Over 600 volts, nominal conductors having nonshielded insulation and operating at different voltage levels are not permitted to be installed in the same cable, raceway, or enclosure.

Reason

The move collects all the applicable rules into one location. The revision clarifies the rule and makes it easier to locate information in the *Code*. The new statement exemplifies the increased attention given to manholes in the 1999 *Code*.

NONMETALLIC-SHEATHED CABLE   300-4(b)(1)

PROTECTION AGAINST PHYSICAL DAMAGE   *REVISED*

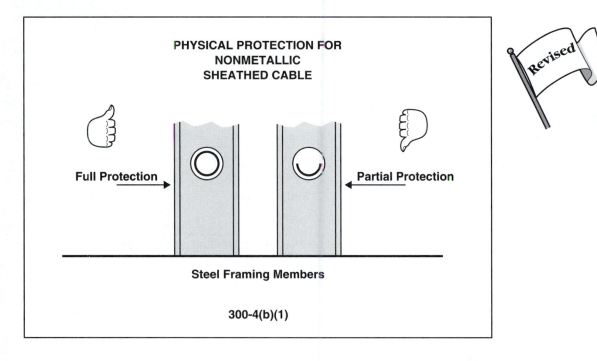

PHYSICAL PROTECTION FOR NONMETALLIC SHEATHED CABLE

Full Protection

Partial Protection

Steel Framing Members

300-4(b)(1)

REVISION

Conductors in cables, nonmetallic cable, or nonmetallic tubing installed in holes in framing members require protection from physical damage by ***holes in metal members with sharp edges,*** by the use of bushings or grommets.

Reason

The use of a U- or V-type grommet or bushing to protect the nonmetallic-sheathed cable installation in a manufactured or a punched hole in a metal framing member leaves sharp edges uncovered. There is record of a severe electrical shock resulting from such an installation.

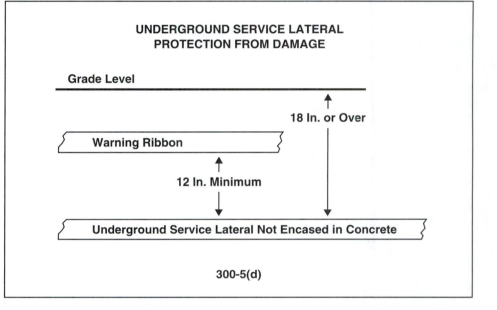

UNDERGROUND SERVICE LATERAL
PROTECTION FROM DAMAGE

Grade Level

18 In. or Over

Warning Ribbon

12 In. Minimum

Underground Service Lateral Not Encased in Concrete

300-5(d)

NEW RULE

Add to the existing section: *When service lateral conductors are buried 18 inches or more and are not encased in concrete, a warning ribbon is required to be installed a minimum of 12 inches above the service lateral.*

Reason

This revision lessens the hazard of digging into unprotected conductors.

Comment

For a dwelling unit, a direct buried-service lateral cable would be at least 18 inches deep with the warning ribbon 12 inches above that. Thus anyone digging would hit the warning ribbon at 6 inches. The regulation does not specify markings for the ribbon. However a 6 inch wide red ribbon marked "CAUTION BURIED ELECTRIC LINE" is available.

SUPPORTING ELECTRICAL WIRING   300-11(a)(1)

ABOVE SUSPENDED CEILINGS   *REVISED*

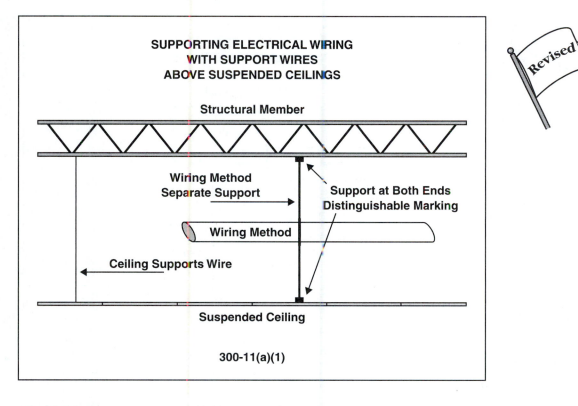

SUPPORTING ELECTRICAL WIRING
WITH SUPPORT WIRES
ABOVE SUSPENDED CEILINGS

Structural Member

Wiring Method
Separate Support

Support at Both Ends
Distinguishable Marking

Wiring Method

Ceiling Supports Wire

Suspended Ceiling

300-11(a)(1)

BASIC RULE

The basic rule requires wiring methods installed in the cavity above a fire-rated ceiling assembly to be securely supported

REVISION

A rigid support method for the support of the wiring method in the cavity above the fire-rated ceiling is permitted.

Support wires, independent of the ceiling support wires for the support of a wiring method above a suspended ceiling are permitted, provided they are secured at both ends:

(1) Fire-rated: *Support wires are to be distinguishable from ceiling support wires by color, tagging, or other effective means.*

(2) Nonfire-rated: The distinguishable requirement is not required for nonfire-rated ceiling or roof/ceiling assembly.

Reason

The revision is made to clarify that independent support wires are permitted when secured at both ends and when they have a distinguishable marking. An independent wire, not supported at both ends, is not considered a secure support.

Comment

Under fault conditions, the fault current will have a whipping effect on the containing raceway. When independent support wires are used, the raceway is not securely fastened in place.

FREE CONDUCTOR LENGTH   *NEW*

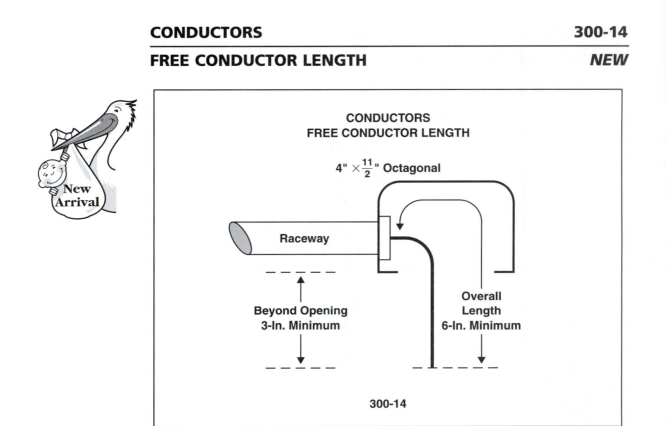

NEW RULE

At all outlet, junctions, or switch point boxes, a minimum of 6 inches of free conductor is required for splice, fixture, or device connection. The 6-inch measurement is from the point the conductor emerges from the raceway or cable. Where the box opening is less than 8 inches in any dimension, a minimum of 3 inches must extend beyond the box opening.

Reason

This new rule settles the controversy over the point from which to measure the required free conductor. It is the result of the Correlating Committee's request of a task force to come up with a firm measurement.

Comment

The original rule required 6 inches of free conductor but did not indicate the point to measure. Some localities measured from the outer edge of the box; others from the point the conductor emerges from the raceway or cable.

BOXES, CONDUIT BODIES AND FITTING 300-15

WHERE REQUIRED *REVISED*

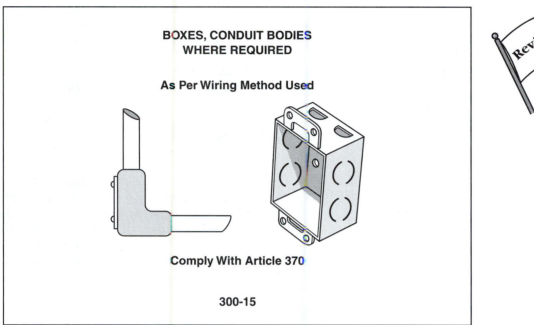

BOXES, CONDUIT BODIES
WHERE REQUIRED

As Per Wiring Method Used

Comply With Article 370

300-15

1996 Wording

Delete: 300-15(a) ~~A box or conduit body complying with sections 370-16 and 370-28 shall be installed at each conductor splice connection point, outlet, switch point, junction point, or pull point for the connection of conduit, electrical metallic tubing, surface raceway or other raceway.~~

Comment

Section 300-15 Boxes, Conduit Bodies, or Fitting—Where Required is completely rearranged with ten exceptions made into permissive rules and four new permissive rules added.

REVISION

Cross Check

1996	1999
300-15 Boxes, Conduit Bodies, or Fitting—Where Required	300-15 Boxes, Conduit Bodies, or Fitting—Where Required
300-15(b) Box Only	(a) Box or Conduit Body
300-15(d) Equipment	(b) Equipment
300-15(b) *Exception No. 1*	(c) Protection
Exception No. 3	(d) Type MI Cable
Exception No. 4	(e) Integral Enclosure
FPN	FPN
Exception No. 8	(f) Fitting
Exception No. 9	(g) Buried Conductors
Exception No. 2	(h) Insulated Devices
New	(i) Enclosure

300-15(a)	*Exception No. 2*	(j)	Fixtures
	New	(k)	Embedded
300-15(b)	*Exception No. 5*	(l)	Manufactured Wiring Systems
	Exception No. 7	(c)	Closed Loop
	New	(n)	Manholes

NEW PERMISSIVE RULES

Splices are permitted to be made without a box in:
(i) gutter of panels or motor control centers where there is sufficient
*   room*
(k) embedded heating cable
(n) manholes that are accessible to qualified persons only

Reason

The basic rule requiring a box at each "conductor splice" could be interpreted as requiring a box for a splice in an auxiliary gutter or a wireway. The reference to the various sections of Article 370 are deleted, and only Article 370 is mentioned. When a box or conduit body is installed, all applicable parts of Article 370 should be fulfilled. The rearrangement is to make the text more user-friendly. The new locations where boxes are not required are coordination with other parts of the *Code*.

RACEWAY INSTALLATION 300-18(a)

PREWIRED RACEWAYS *NEW*

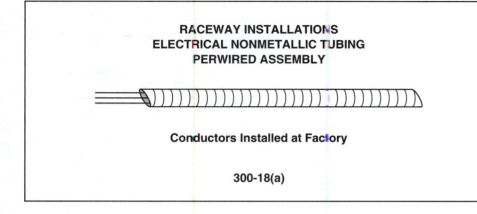

**RACEWAY INSTALLATIONS
ELECTRICAL NONMETALLIC TUBING
PERWIRED ASSEMBLY**

Conductors Installed at Factory

300-18(a)

REVISION

Section 300-18 Raceway Installations, is revised, and the four exceptions are reworded and made a part of the text.

BASIC RULE

The basic rule requires all raceways and boxes to be completely installed before conductors are installed. One new permissive rule is added.

NEW PERMISSIVE RULE

Prewired raceway assemblies are now permitted provided they are specifically designed to be or otherwise specifically permitted to be prewired in this Code.

Reason

Electrical nonmetallic tubing and liquidtight flexible nonmetallic conduit are now permitted to be prewired, and this permissive rule coordinates with that.

User Friendly Code Rule Movers
Moving From Here to There

Revised

OVER 600 VOLTS NOMINAL GENERAL INSTALLATION
MOVE FROM ARTICLE 710 ⟶ TO ARTICLE 300 PART B

Electrical Metallic Tubing | 5000 Volt Conductors

300-37

MOVED

The general requirement methods for installation of conductors over 600 volts, nominal is moved from Article 710, which is now deleted, and is coordinated into the existing Part B. of Article 300, "Conductors Over 600 Volts, Nominal." *Electrical metallic tubing is now recognized as a permitted wiring method for over 600 volts, nominal above ground.*

Cross Check

1996	1999
Part B. Requirements Over 600 Volts Nominal	Part B. Requirements Over 600 Volts Nominal
300-31 Cover Required *Exception No. 1* Moved to 300-3(c)(2)d *Exception No. 2* Moved to 300-3(c)(2)e	300-31 Cover Required
300-32 Conductors of Different Systems [Delete text and move to 300-3(c)] 300-34 Conductor Bending Radius 300-35 Protection Against Induction Heating	300-32 Conductors of Different Systems Reference 300-3(c)(2) 300-34 Conductor Bending Radius 300-35 Protection Against Induction Heating
300-36 Grounding Delete Covered in Article 250 300-37 Underground Installations Delete Referenced Article 710	
710-4 Wiring Methods (a) Aboveground Conductors 710-4(c) Busbars (Aluminum or Copper) 710-5 Braid-Covered Insulated Conductors—Open Installation 710-6 Insulation Shielding	300-37 Aboveground Wiring Methods Included in text 300-39 Braid-Covered Insulated Conductors—Open Installation 300-40 Insulation Shielding
710-7 Grounding Delete Covered in Article 250 710-8 Moisture or Mechanical Protection for Metal-Sheathed Cables	300-42 Moisture or Mechanical Protection for Metal-Sheathed Cables

700-4(b) Underground Conductors
    Exception No. 1
    and *Exception No. 2*
 	
700-4(b)   Part of text
700-4(b)(1) Protection from Damage
    (2) Splices
      Exception
    (3) Backfill
    (4) Raceway Seal
	300-5 Underground Installation
	(a) General
	    (1) Shielded Cables and
	        Nonshielded Cables in Metal-
	        Sheathed Cable Assemblies
	    (2) Other Nonshielded Cables
	(b) Protection from Damage
	(c) Splices
	    Exception
	(d) Backfill
	(f) Raceway Seal

Table 710-4(b)
    Exception No. 1
    Exception No. 2
    Exception No. 3
    Exception No. 4
    Exception No. 5
    Exception No. 6
	Table 300-50
	    Exception No. 1
	    Exception No. 2
	    Exception No. 3
	    Exception No. 4
	    Exception No. 5
	    Exception No. 6

NONMETALLIC-SHEATHED CABLE PERMITTED *REVISED*

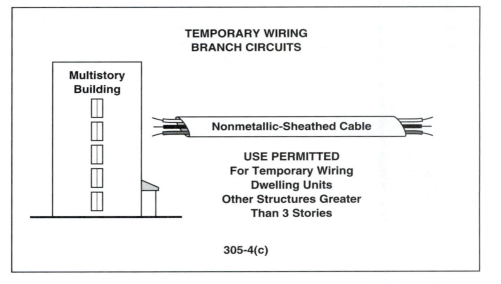

TEMPORARY WIRING BRANCH CIRCUITS

Multistory Building

Nonmetallic-Sheathed Cable

USE PERMITTED
For Temporary Wiring
Dwelling Units
Other Structures Greater
Than 3 Stories

305-4(c)

REVISION

305-4(c) Branch Circuits Nonmetallic-sheathed cable types NM and NMC are permitted to be used for temporary wiring in dwelling units and other structures without height limitations.

Reason

The basic rule limits the use of nonmetallic-sheathed cable to three stories. This revision makes clear that types NM and NMC can be used for temporary wiring in multistory buildings.

ADD NEW SUBSECTION

305-4(j) Where flexible cords and cables are used for temporary wiring, they are required to be supported at intervals that ensure that the cords or cables will be protected from physical damage. Support is permitted to be in the form of staples, cable ties, straps, or similar type fittings so designed as not to damage the cord or cable assembly.

Reason

Section 305-4(a) and (b) permits hard or extra-hard usage cords or cables to be used for temporary wiring, and the support of these cords or cables is not directly indicated.

305-6 Ground-Fault Protection for Personnel

The titles for the subsections of the section are deleted and new titles introduced. Despite rewording, there is no change in the intent of the rule.

Cross Check

1996	1999

305-6 Ground-Fault Protection for Personnel
  (a) Ground-Fault Circuit Interrupters
  (b) Assured Equipment Grounding
	305-6 Ground-Fault Protection for Personnel
	  (a) Receptacle Outlets
	  (b) Use of Other Outlets
	    (1) GFCI for personnel
	    (2) Assured system

ARTICLE 310

CONDUCTORS FOR GENERAL WIRING

310-4 Conductors in Parallel

Reword and move the following from the FPN into the text:

It is not necessary for conductors of one phase, neutral, or grounded circuit conductor to have the same physical characteristic as those of another phase, neutral, or grounded circuit conductor to achieve balance.

Comment

The former FPN was being applied as a rule in the field because it indicated part of the intent of the rule. Where possible, such FPNs are being made a part of the enforceable text.

310-5 Minimum Size of Conductors

The basic rule is that the minimum size for a conductor is as per Section 310-15.

New exception: *Exception No. 9: Control and instrumentation circuits as permitted by Section 727-4.*

Reason

Article 727 Instrument Tray Cable is introduced into the *Code* in 1996 as a separate article, so coordination of conductor size is necessary.

310-8 Location

Section 310-8 ~~Wet Locations~~ of the 1996 *Code* is deleted and a new and more identifiable outline is introduced. Specific insulations for specific installations are listed. The following illustrates the new subtitles and comments briefly on the content.

310-8 Location

(a) Dry Location	All *Code* recognized insulations
(b) Dry and Damp Locations	22 insulation types listed
(c) Wet Locations	
(1)	Special covering
(2)	13 insulation types listed
(3)	Listed for use
(d) Location Exposed to Direct Sunlight	

Reason

The rearrangement and addition of the types of insulations make this section easier to apply. The new subsection (d) is added because the *Code* refers to direct sunlight several times.

310-11 Marking

A new requirement is added in **310-11(a)(5)**.

(a) Required Information
(5) Manufactured cable assemblies where the neutral conductor is smaller than the ungrounded conductors shall be so marked.

Comment

This marking requirement is moved from 210-19(a) because it refers to cable markings. By relocating it to this section, it now applies to feeders and branch circuits. For example, a cable with the grounded conductor (neutral) could be smaller than the ungrounded conductors for a 120/240-volt dryer.

310-12 Conductor Identification

Delete **(a)** and **(b)** of the 1996 *Code* and all their exceptions and substitute the following:
(a) Grounded Conductors *As per Section 200-6.*
(b) Equipment Grounding Conductor *As per Section 250-119.*

Reason

The information presented here was redundant to information already presented in the *Code*. The sections indicated are the logical place for the rule.

310-15 Ampacities for Conductors Rated 0–2000 Volts

This section is completely rearranged, the title is revised, and the notes to ampacity tables of 0–2000 volts, with their exceptions, are deleted and made a part of the text of Section 310-15.

Cross Check

1996	1999
310-15 Ampacities	**310-15 Ampacities for Conductors Rated 0–2000 Volts** (title only)
	(a) General (title only)
New title and text	**(1) Tables or Engineering Supervision** (choice)
310-15(a) (FPN) Voltage Drop	(FPN No. 1)
Table Note 7. Type MTW Machine Tool Wire	(FPN No. 2)
310-15(c) Selection of Ampacity	**(2) Selection of Ampacity**
Exception	*Exception*
(FPN)	(FPN)
310-15(a) General As per T310-16 > 310-19	**(b) Tables** As per T310-16 > 310-19 As modified by 1–7 below.
310-15(a) General (FPN)	(FPN)
310-15(a) 1.	**(1)** (Temperature)
2.	**(2)** (Overcurrent)
3.	**(3)** (Product listing)
4.	**(4)** (Industry Safety)
	(1) General See T310-13
Note 8 Adjustment Factor	**(2) Adjustment Factors**
(a) More Than 3 Current-Carrying in a Raceway or Cable	**(a) More Than 3 Current-Carrying in a Raceway or Cable**
Table	**Table 310-15(b)(2)**
(FPN)	(FPN)
Exception No. 1	*Exception No. 1*
Exception No. 2	*Exception No. 2*
Exception No. 3	*Exception No. 3*
Exception No. 4	*Exception No. 4*

(b) More than One Conduit, Tube, or Raceway	**(b) More than One Conduit, Tube, or Raceway**
Note 5 Bare or Covered Conductors	**(3) Bare or Covered Conductors**
Note 10 Neutral Conductor	**(4) Neutral Conductor**
a.	(a)
b.	(b)
c.	(c)
Note 11 Grounding or Bonding Conductor	**(5) Grounding or Bonding Conductor**
Note 3 120/240 Volts, 3-Wire, Single-Phase Dwelling Service and Feeder	**(6) 120/240 Volts, 3-Wire, Single-Phase Dwelling Service and Feeder**
	Reworded with some revision
Table	**Table 310-15(b)(6)**
Note 6. Mineral-Insulated Metal-Sheathed Cable	**(7) Mineral-Insulated Metal-Sheathed Cable**
310-15(c) Engineering Supervision	**(c) Engineering Supervision**

Tables 310-85 and 310-86

Ampacity of Three Triplexed Single-Insulated Copper Conductors Directly Buried in Earth Based on Ambient Earth Temperature of 20° C (68° F). Arrangement per Figure 310-60, 100% Load Factor, Thermal Resistance RHO of 90, Conductor Temperature 90°C (194°F) and 105°C (221°F). These two tables are revised to list the ampacity for both Type MV-90 and MV-105 cables because Type MV 105 was previously left off the table. Table 310-1 of the 1996 *Code* has been renumbered Table 310-60 and moved ahead of the tables it applies to instead of following them.

User Friendly Code Rule Movers
Moving From Here to There

CONDUCTOR AMPACITY

Table 310-20   Ampacities of Two or Three Single Insulated Conductors, Rated 0 Through 2000 Volts, Supported on a Messenger, Based on Ambient Air Temperature of 40°C (104°F)

Size	Temperature Rating of Conductor, See Table 310-13.				Size
	75°C (167°F)	90°C (194°F)	75°C (167°F)	90°C (194°F)	
AWG kcmil	TYPES RH, RHW THHW, THW, THWN, XHHW, ZW	TYPES THHN, THHW, THW-2, THWN-2, RHH, RWH-2, USE-2, XHHW, XHHW-2, ZW-2	TYPES RH, RHW, THW, THWN, THHW, XHHW	TYPES THHN, THHW, RHH, XHHW, RHW-2, XHHW-2 THW-2, THWN-2, USE-2, ZW-2	AWG kcmil
	COPPER		ALUMINUM OR COPPER-CLAD ALUMINUM		
8	57	66	44	51	8
6	76	89	59	69	6
4	101	117	78	91	4
3	118	138	92	107	3
2	135	158	106	123	2
1	158	185	123	144	1
1/0	183	214	143	167	1/0
2/0	212	247	165	193	2/0
3/0	245	287	192	224	3/0
4/0	287	335	224	262	4/0
250	320	374	251	292	250
300	359	419	282	328	300
350	397	464	312	364	350
400	430	503	339	395	400
500	496	580	392	458	500

TABLE 310-20

MOVE

Move Table B310-2 from Appendix A to Article 310 and make it *Table 310-20, Ampacity of Two or Three Single Insulated Conductors, Rated 0 through 2000 Volts, Supported on a Messenger, Based on Ambient Air Temperature of 40°C (140°F).*

Reason

The existing tables in Article 310 of the *Code* did not really cover this particular cable assembly installation.

Comment

Moving this table out of Appendix B takes it from "under engineering supervision," and the journeyman electrician in the field and others can now use the table. In the move, only the table number is changed; the values remain the same.

CONDUCTORS

AMPACITY

TABLE 310-21

MOVED

User Friendly Code Rule Movers
Moving From Here to There

Conductor AmPAcity

Table 310-21 Ampacities for Bare or Covered Conductors
Based on 40°C (104°F) Ambient, 80°C (176°F) Total Conductor Temperature,
2 Feet (610 mm) per Second Wind Velocity

Bare Copper Conductors		Covered Copper Conductors	
AWG kcmil	AMPS	AWG kcmil	AMPS
8	98	8	103
6	124	6	130
4	155	4	163
2	209	2	219
1/0	282	1/0	297
2/0	329	2/0	344
3/0	382	3/0	401
4/0	444	4/0	466
250	494	250	519
300	556	300	584
500	773	500	812
750	1000	750	1050
1000	1193	1000	1253

Bare AAC Aluminum Conductors		Covered AAC Aluminum Conductors	
AWG kcmil	AMPS	AWG kcmil	AMPS
8	76	8	80
6	96	6	101
4	121	4	127
2	163	2	171
1/0	220	1/0	231
2/0	255	2/0	268
3/0	297	3/0	312
4/0	346	4/0	364
266.8	403	266.8	423
336.4	468	336.4	492
397.5	522	397.5	548
477.0	588	477.0	617
556.5	650	556.5	682
636.0	709	636.0	744
795.0	819	795.5	860
954.0	920		
1033.5	968	1033.5	1017
1272	1103	1272	1201
1590	1267	1590	1381
2000	1454	2000	1527

TABLE 310-20

MOVE

Move Table B310-4 from Appendix A to Article 310 and re-identify it as *Table 310-21 Ampacity for Bare or Covered Conductors Based on 40° (104°F) Ambient, 80°C (176°F) Total Conductor Temperature, 2 Feet (610 mm) per Second Wind Velocity.*

Moving this table from the Appendix eliminates the need for the column "Bare and Covered Conductors" in Table 310-19, and it is deleted.

Reason

This move gives bare and covered conductors a more applicable separate ampacity table.

Comment

Many standard conductors such as No. 12, 14, 10, 3, 1, 350 kcmil, 400 kcmil, 600 kcmil, 800 kcmil, 900 kcmil, and 1500 kcmil are not listed in this table. Moving the table does not change the values. This new table applies to bare or covered conductors installed outside and suspended in air.

Section 310-15(b)(3) indicates that the ampacity of a bare or covered conductor installed with insulated conductor is the ampacity of the conductors it is installed with. For example, where a bare conductor is installed with the service-entrance conductors, the ampacity of the bare conductor is based on the insulation of the other conductors in the same raceway.

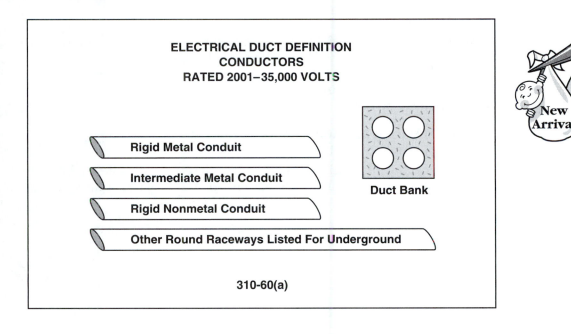

NEW DEFINITION

Electrical ducts shall include any of the electrical conduits recognized in Chapter 3 as suitable for underground and other raceways round in cross section, listed for underground use, embedded in earth or concrete.

The term "Thermal Resistivity" is also defined here because it is often used in conjunction with the installation of the higher voltage conductors in ducts.

As used in the *Code*, the term *"Thermal Resistivity" means the heat transfer capability through a substance by conduction. It is the reciprocal of thermal conductivity, is designated rho, and is expressed in the units °C-cm/watt.*

A new *Section 310-60(b) permits the ampacity of conductors 2001–35,000 volts to be calculated under engineering supervision in accordance with the requirements list in this section.*

Reason

The task group working with this area of the *Code* felt there was a need for these definitions.

Comment

The definition for electrical ducts relates to Figure 310-60 of the *Code*, which illustrates underground duct banks. Section 310-60 starts the higher voltages. This is the first of several changes applicable to installations with conductors rated 2001 volts or higher.

ARTICLE 318

CABLE TRAYS

318-3 Uses Permitted

(a) Wiring Methods. The following wiring methods are an addition to the wiring methods permitted to be installed in cable trays.
Fire alarm cables, Article 760.
Multipurpose communications cable.
Power-limited tray cable.

318-3(b)(1) Wiring Methods Single Conductor in Industrial Establishments

The two exceptions following *318-3(b)(1)* are deleted and made a part of the text without a change in intent. Exception No. 1 referred to welding cables, and Exception No. 2 referred to the grounding conductor.

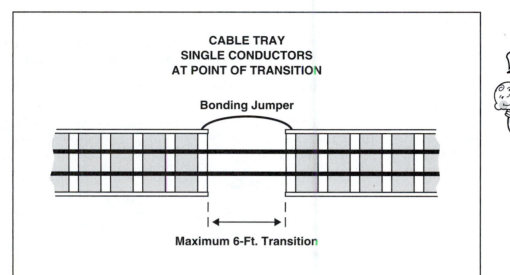

CABLE TRAY
SINGLE CONDUCTORS
AT POINT OF TRANSITION

Bonding Jumper

|◄──────►|
Maximum 6-Ft. Transition

318-6(a)

NEW RULE

Where cable trays support individual conductor, the conductors are permitted to pass from one cable tray to the another, or to a raceway, or to equipment where all the following conditions are met:
1. The transition space does not exceed 6 ft.
2. Conductors are secured to cable tray at point of transition.
3. The conductors are protected by guarding or location.
4. Bonding is required at point of transition.

Reason
This new rule addresses the problem of transition from cable tray to cable tray or from cable tray to raceway or equipment.

318-6 Installation
(f) Cables Rated at Over 600 Volts The two exceptions following Section 318-6(f) are deleted and made a part of the text without a change in intent. Exception No. 1 concerned the use of barriers, and Exception No. 2 concerned Type MC cable.

318-6 Installation
(j) Raceway Cables and Outlet Boxes Supported from Cable Trays In industrial installations with proper supervision, the cable tray is now permitted to support outlet boxes provided they meet the mounting requirements of Article 370.

Reason
If the cable tray can support the wiring method, it can also support the outlet box.

132

318-8 Cable Installation

(d) Connected in Parallel The exception is deleted and made a part of the text. The exception concerned the installation of triplex cable in cable trays.

318-8 Cable Installation

(e) Single Conductors The exception is deleted and made a part of the text. The exception concerned grouping the conductors to protect against induction.

318-11 Ampacity of Cables, Rated 2000 Volts or Less, in Cable Trays

(a) Multiconductor Cables Delete the two exceptions and rearrange the text into a readable outline. The exceptions are included in the rearrangement with no change in the intent of the rule.

Cross Check

1996	1999
318-11(a) Multiconductor Cables	318-(a) Multiconductor Cables
	Part of text
318-11(a) Multiconductor Cables	(1) Part of Text
Exception No. 1	(2)
Exception No. 2	(3)
(FPN)	(FPN)

CABLE TRAY   318-11(b)(4)

SPACING CABLES 2000 VOLTS OR LESS   *REVISED*

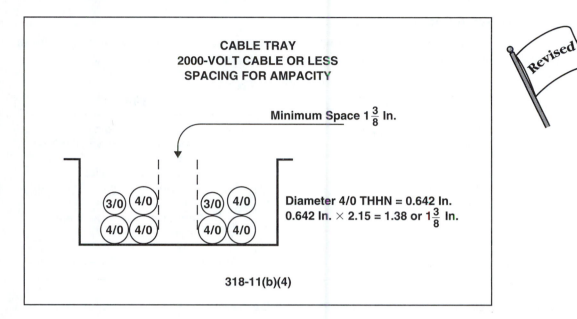

CABLE TRAY
2000-VOLT CABLE OR LESS
SPACING FOR AMPACITY

Minimum Space $1\frac{3}{8}$ In.

Diameter 4/0 THHN = 0.642 In.
0.642 In. $\times$ 2.15 = 1.38 or $1\frac{3}{8}$ In.

318-11(b)(4)

REVISION

Where single conductors are installed in a triangular or square configuration in uncovered cable trays, with a maintained *free air* space of not less than 2.15 times the ~~one conductor~~ diameter (2.15 × O.D.) *of the largest conductor contained within the configuration and adjacent conductor configuration of cables,*—etc.

Reason

This revision clarifies the free-air requirement between cables installed in cable trays.

Revision

318-13(b)(3) Ampacity of Type MV and Type MC Cable (2001 Volts or Over) in Cable Tray A similar revision is made for cable 2001 volts or over when cables are installed in a configuration.

ARTICLE 320

OPEN WIRING ON INSULATORS

320-6 Conductor Support

This section is rearranged with the exceptions reworded as permissive rules in the text as illustrated in the following cross check.

Cross Check

1996	1999
320-6 Conductor Support	**320-6 Conductor Support**
Part of text	**(a) Conductors Sizes Smaller Than No. 8**
Part of text	**(1)** 12 in. tap or splice
Part of text	**(2)** 12 in. dead enc
Part of text	**(3)** 4 1/2 ft intervals
Exception No. 1	**(b) Conductors Sizes No. 8 and Larger**
Exception No. 2	Included in text
Exception No. 3	**(c) Industrial Establishments**

320-14 Protection from Physical Damage

Delete last sentence of the text. ~~The conductors passing through metal enclosures shall be so grouped that current in both directions is approximately equal.~~

Reason

This requirement is redundant to Section 320-2, which requires the installation of open wiring on insulators to be installed in accord with Article 300. Section 300-20 addresses induced currents.

ARTICLE 330

MINERAL-INSULATED METAL-SHEATHED CABLE

330-4 Used Not Permitted

Delete the exception and include it in the text.

330-12 Supports

Delete the two exceptions and rearrange text with easier-to-identify subdivisions. The exceptions are included in the rearrangement and rewording with no change in the intent of the rule.

Cross Check

1996	1999
330-12 Supports	**330-12 Supports** Support as per below:
330-12 Supports	(1) Basic support
Exception No. 1	*Exception*
Exception No. 2	(2) Cable tray

ELECTRICAL NONMETALLIC TUBING 331-3(6)

PERMITTED USE *NEW*

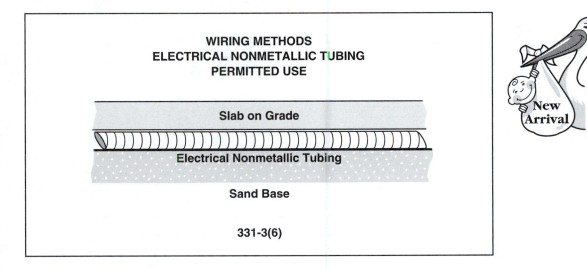

NEW RULE

Electrical nonmetallic tubing is permitted to be installed below a slab on grade when placed on a sand or approved screenings base.

Reason

Damage will not occur to the ENT when placed onto suitable material under a concrete slab because it is recognized for encasement in concrete.

```
WIRING METHODS
ELECTRICAL
NONMETALLIC TUBING


Electric Nonmetallic Tubing Prewired Assembly


331-3(8)
```

NEW RULE

Electrical nonmetallic tubing is permitted to be manufactured as a listed prewired assembly in sizes 1/2 inch through 1 inch. Section 331-15 The enclosed conductors are required to be marked at each end of the assembly and on either the coil, carton, or reel. The tubing must be provided in continuous lengths shipped in a coil, reel, or carton.

Reason

This revision recognizes a new product on the market.

Comment

In the past, the basic *Code* rule has required that a raceway for general use be completely installed before the conductors are installed, Section 310-18. The basic rule is there for the protection of the conductors. This is a departure from that rule and now opens up a new era of prewired raceways that can be cut and fit in the field.

ARTICLE 333

ARMORED CABLE

333-7 Support

Delete the five exceptions and rearrange the text into an outline. The exceptions are included in the rearrangement with no change in the intent of the rule.

Cross Check

1996	1999
333-7 Support	333-7 Support
Exception No. 5	(a) Horizontal Runs
	(b) Unsupported Cables
Exception No. 1	(1) fished
Exception No. 2	(2) flexibility
Exception No. 3	(3) fixture whip
Exception No. 4	(c) Cable Tray Installations

333-11 Exposed Work

Delete Exception No. 2 and include it in the text. Exception No. 2 refers to installations on the underside of joists.

Delete ~~*Exception No.1: Lengths of not more than 24 in. (610 mm) at terminals where flexibility is necessary.*~~

~~*Exception No. 3: Lengths of not more than 6 ft (1.83 m) from an outlet for connection within an accessible ceiling to lighting fixtures or other equipment.*~~

Reason

The two exceptions are deleted because they are redundant to Section 333-7. To make the *Code* more user-friendly, redundant rules are being deleted.

ARTICLE 334

METAL-CLAD CABLE

334-3 Use Permitted

This long paragraph listing the many uses of Type AC Cable has been rearranged into an easy-to-read outline. Add: ***Metal-clad cable is now permitted for installation as permitted in Article 505.***

Delete: *Exception: See Section 501-4(b), Exception*

Reason

The revision coordinates with a permitted use in Article 505.

334-4 Uses Not Permitted

Delete the exception and include it in the text. Where the metallic jacket is identified for installation in a corrosive location, it is permitted to be installed there.

334-10 Installation

(a) Support This section is rearranged with the exceptions reworded as permissive rules in the text as illustrated in the following cross check.

Cross Check

1996	1999
334-10 Installation	334-10 Installation
(a) Support	(a) Supported Cables
Exception No. 3	(1) Horizontal Runs
	(2) At Terminations
Exception No. 1	(b) Unsupported Cables
(b) Cable Tray	(c) Cable Tray
(c) Direct Burial	(d) Direct Burial
(d) Installed as Service-Entrance Cable	(e) Installed as Service-Entrance Cable
(e) Installed Outside of Building or as Aerial Cable	(f) Installed Outside of Building or as Aerial Cable
(f) Through or Parallel to Joints, Studs, and Rafters	(g) Through or Parallel to Joints, Studs, and Rafters
(g) In Accesible Attics	(h) In Accesible Attics

334-24 Marking

Delete: *Cable that are flame-retardant and have limited-smoke characteristics shall be permitted to be identified with the suffix LS.* This statement is already covered in Section 310-11(d), Optional Markings.

ARTICLE 336

NONMETALLIC-SHEATHED CABLE

336-6 Installation—Exposed Work (b) Protection from Physical Damage

Two additional raceways are recognized for use to protect conductors: *listed surface metal raceway and nonmetallic raceway.*

336-25 Device with Integral Enclosure

This new section recognizes the use of nonmetallic-sheathed cable with a special wiring device with integral enclosure, is identified for such use. It coordinates with Section 300-15(e) that also permits it.

336-26 Ampacity

This section is made by moving part of the text of 336-26 without changing the intent of the rule. The following shows previous locations.

Cross Check

1996	1999
None 336-30(b) Move part of text 336-30(b) *Exception*	**336-26 Ampacity** Text consists of: 60°C ampacity for conductors Applying derating factors

336-31 Marking

Delete: ~~Cables that are flame-retardant and have limited-smoke characteristics shall be permitted to be identified with the suffix LS.~~ This statement is redundant because it is covered in Section 310-11(d), Optional Markings.

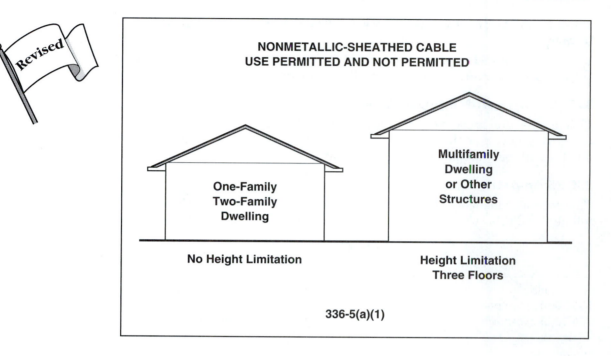

NONMETALLIC-SHEATHED CABLE
USE PERMITTED AND NOT PERMITTED

One-Family
Two-Family
Dwelling

Multifamily
Dwelling
or Other
Structures

No Height Limitation

Height Limitation
Three Floors

336-5(a)(1)

REVISION

336-5(a)(1) In any ***multifamily*** dwelling or ***other*** structure exceeding three floors above grade.

Delete ~~Exception permitting the renovation of attic, vehicle parking or storage space created into habitable room.~~

Reason

There was no reason to restrict the use of nonmetallic-sheathed cable to three floors.

Comment

With the insertion of the word "multifamily" into the rule, it now permits a single family dwelling to use nonmetallic-sheathed cable in a one- or two-family dwelling without restriction to height. Without the height restriction, there was no need for the exception. Then, how many mulitstory single- or two-family dwellings do you see that are over three stories high?

NONMETALLIC-SHEATHED CABLE   336-18 Exception No. 3

SUPPORTS   *REVISED*

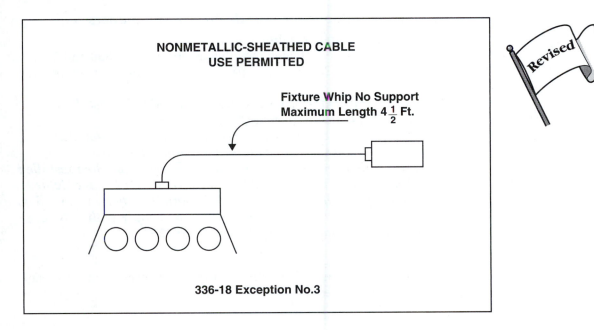

REVISION

Nonmetallic-sheathed cable is now permitted for fixture whips. The whip is limited in length to 4 ¹/₂ ft without being secured within 12 inches from the lighting fixture.

Reason

Section 410-14(a) permits the use of nonmetallic-sheathed cable for lighting fixtures. This revision coordinates with Section 410-14(a). The 4 1/2-ft limitation is used rather than the 6-ft limitation, because that is the maximum allowed between supports for nonmetallic-sheathed cable.

REVISION 336-18

Change the words "Two conductor cables" to read *"Flat cables."*

Reason

The change prevents the stapling of any flat conductor cable on the edge, regardless of the number of conductors.

Delete the FPN following 336-18 and make it a part of the text. The FPN referenced Article 370, Boxes.

Delete the two exceptions and reword them as part of the text with no change in intent.

ARTICLE 338

SERVICE-ENTRANCE CABLE

338-4 Installation Methods for Branch circuits and feeders

This section is revised into two subsections.

(a) Interior Installations. No change installed as per Articles 336 and 300.

(b) Interior Installations. This is a new part with the following new requirements.

1. *When installed as exterior wiring, it is required to be installed as per Article 225, Outside Branch Circuits and Feeders.*
2. *The cable is required to be supported the same as nonmetallic-sheathed cable unless used as messenger supported wiring.*
3. *Type USE Cable is permitted to be used underground when installed as per Article 339, Underground Branch-Circuit and Feeder Cable.*
4. *Multiconductor service-entrance cable is permitted to be installed as messenger support wiring, and is required to comply with Article 321, Messenger Support Wiring, and Article 225.*

Reason

This revision clarifies the use of the different types of service-entrance cable and gives specific guidelines for their installation.

ARTICLE 339

UNDERGROUND FEEDER AND BRANCH-CIRCUIT CABLE

339-1(b) Marking

Delete: ~~The cable shall have a distinctive marking on the exterior for its entire length that specifies the cable type.~~ This is redundant. It is already covered in Section 310-11(d), Optional Markings.

339-3(a) Uses Permitted

Delete the exception and reword it as parts (5) and (6) of the text without changing the intent of the rule.

339-3(b) Uses Not Permitted

Add: *(10) where subject to physical damage.*

ARTICLE 340

POWER AND CONTROL TRAY CABLE

340-3 Construction

The two exceptions following Section 340-3 are reworded into permissive rules and the section is rearranged as follows.

Cross Check

1996	1999
340-3 Construction	340-3 Construction
Part of text	(a) Wet Locations
Exception No. 1	(b) Fire Alarms
Exception No. 2	(c) Thermocouple Circuits
Part of text	(d) Class I Circuit Conductors

340-4 Uses Permitted

A long paragraph consisting of (1), (2), (3), (4), and (5) is separated out. The exception following Section 340-5, Uses Not Permitted, is deleted, reworded, and made a part of the text (6) of Section 340-4. The rule governs the installation of power and control cable tray cable in lengths not exceeding 50 feet in an industrial establishment.

340-5 Uses Not Permitted

A long paragraph consisting of (1), (2), (3), and (4) is separated out.

340-6 Marking

Delete: ~~Cable that are flame-retardant and have limited-smoke characteristics shall be permitted to be identified with the suffix LS.~~ This statement is redundant because it is already covered in Section 310-11(d), Optional Markings.

LIQUIDTIGHT FLEXIBLE NONMETALLIC CONDUIT

PERMITTED USE

351-23(a)(6)

NEW

LIQUIDTIGHT FLEXIBLE
NONMETALLIC CONDUIT PERMITTED USE
PREWIRED

Liquidtight Flexible Nonmetallic Conduit

Listed Smooth Interior, Factory Installed

351-23(a)(6)

New Arrival

NEW RULE

A listed liquidtight flexible nonmetallic conduit with a smooth inner surface and without integral reinforcement within the conduit wall in sizes 1/2 through 1 inch is permitted to be factory prewired in unlimited lengths.

Reason

Liquidtight flexible nonmetallic conduits have been successfully used as fixture whips. Cutting to length in the field is no more difficult than cutting a metal enclosure with conductor, such as MC or AC cable.

351-27 Support—New

The three exceptions following Section 351-27 are reworded into permissive rules, and the section is rearranged as follows.

Cross Check

1996	1999
351-27 Supports	351-27 Security and Supporting
Part of text	(a) Basic support measurements
Exception No. 1	(b) Fished
Exception No. 2	Flexibility
Exception No. 3	Fixture whips
Part of text	(c) Horizontal runs

ARTICLE 352

SURFACE METAL RACEWAYS AND SURFACE NONMETALLIC RACEWAYS

Cross Check

1996	1999
352-1 Uses	**352-1 Uses**
One long paragraph is outlined with no change in intent.	**(a) Permitted**
	(1) Dry locations
	(2) As per 501-4(b)
Exception	**(3)** As per 645(d)(2)
	(b) Not Permitted
	(1) Physical damage
	(2) 300 volts between conductors
	(3) Corrosive vapors
	(4) Hoistways
	(5) Concealed except 352-1(a)(3)

352-7 Splices and Taps

Basic rule: Splices and taps made in surface metal raceways are permitted when the raceway has a cover.

Add the following as a positive statement: ***Taps of Type FC cable installed in surface metal raceways shall be made in accordance with Section 363-10.***

Reason

Section 363-10, Flat Conductor Cable, permits taps to be made in surface metal raceways that are covered.

ARTICLE 362

METAL WIREWAYS AND NONMETALLIC WIREWAYS

362-2 Uses

The section is rearranged with the use of subsections.

362-8(a) Horizontal Supports

Insert a phrase: Where wireways are run horizontally, they are required to be supported at each end, and at intervals not to exceed 5 ft *or for individual lengths longer than 5 ft at each end or joint,* unless listed for other support.

Reason

The previous *Code* did not make clear that an end support would be required for a wireway 5 ft or less.

362-11 Extensions from Wireways

Add a phrase: Extension from wireways shall be made with cord and pendant *in accordance with Section 400-10* or any wiring method that includes a means for equipment grounding.

Reason

This section applies to this type of installation.

Comment

Section 400-10 requires the cord and pendant to be installed without transferring tension to joints or terminations.

362-19 Number of Conductors

Revise the wording in the section: The derating factor specified in Section 310-15(b) shall be applicable to current-carrying conductors ~~at the 20% fill~~ *up to and including the 20% fill permitted.*

Reason

The revision makes clear that the derating factors apply to the 20% fill as well as to lower fills.

METAL WIREWAYS   **362-6**

DEFLECTION OF CONDUCTORS   *NEW*

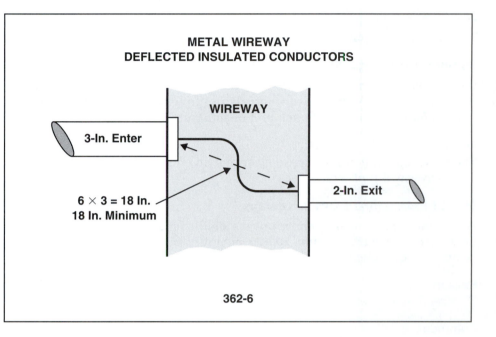

METAL WIREWAY
DEFLECTED INSULATED CONDUCTORS

WIREWAY

3-In. Enter

2-In. Exit

6 × 3 = 18 In.
18 In. Minimum

362-6

NEW RULE

Where the same conductors enter and exit a wireway through a raceway or cable containing insulated conductors No. 4 or larger with a deflection greater than 30 degrees the distance between those raceway or cable entries is required to be not less than six times the trade size diameter of the largest raceway.

Reason

This rule ensures adequate space for bending conductors as they pass through the wireway without damaging the insulation.

Comment

This rule indicates that a wireway used as a pull box requires the same bending space for the conductors as a pull box. Note: A straight through pull is not a deflection.

BUSWAY   **364-6(b)(2)**

THROUGH FLOORS   ***NEW***

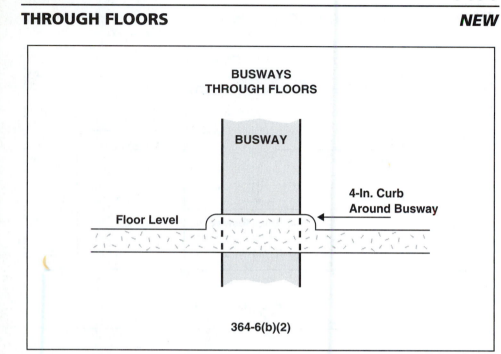

BUSWAYS
THROUGH FLOORS

BUSWAY

4-In. Curb
Around Busway

Floor Level

364-6(b)(2)

New
Arrival

NEW RULE

In other than industrial establishments, a minimum 4-inch curb, within a 12-inch opening is required around floor openings for vertical busway risers penetrating two or more floors to prevent floor level liquids from entering the opening.

Comment

This new rule has the reason for the rule stated within the rule.

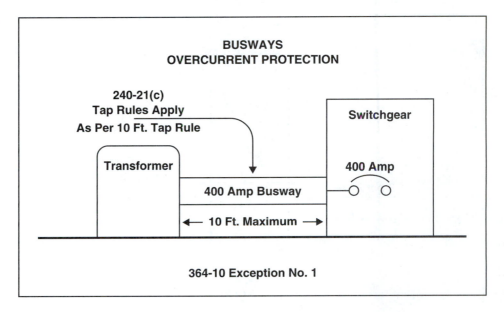

BASIC RULE

364-10 Rating of Overcurrent Protection—Feeders. The busway is required to be protected against overcurrent in accordance with the busway ampacity.

Delete the following from this section: ~~Where the allowable current rating of the Busway does not correspond to a standard ampere rating of the overcurrent device, the next-higher-rated standard overcurrent device shall be permitted, only if this rating does not exceed 800 amperes.~~

Replace with Two New Exceptions

1. Exception No. 1: The applicable provisions of 240-3 shall be permitted.
2. Exception No. 2: Where used as transformer secondary ties, the provisions of Section 450-6(a)(3) shall be permitted.

Reason

The revision clarifies the rule and coordinates it with Article 240, Overcurrent Protection, and Article 450, Transformers.

Comment

By referencing the applicable provisions of Section 240-3 Exception No. 1 covers two issues. First, Section 240-3(b) permits the next size larger overcurrent protective device, provided it does not exceed 800 amperes. Second, 240-3(f) references 240-21(c), making clear that the transformer tap rules listed there can be applied to busways.

Exception No. 2 coordinates with 450-6(a)(3), which sets down the overcurrent protection regulations that apply to busways when they are installed as transformer secondary ties.

BUSWAY—BRANCH CIRCUIT   364-13

OVERCURRENT PROTECTION   *REVISED*

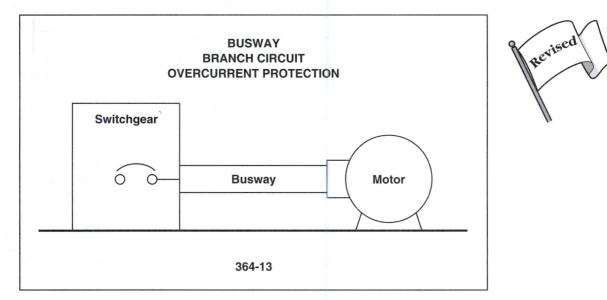

PREVIOUS RULE

A branch circuit busway was required to have overcurrent protection as specified in Article 210.

REVISION

A branch circuit busway is required to have overcurrent protection according to 210-20. Where used as a branch circuit listed in Article 210, 430, or 440, it is required to comply with the requirements for that particular branch circuit.

Reason

The revision clarifies when the busway wiring method is used for various types of branch circuits.

Comment

Section 210-20 brings the busway under the overcurrent protection rule for branch circuit continuous and noncontinuous loads. Article 210 bring the busway under the rules in Article 210 for the installation of the various types of branch circuits. Article 430 is listed to cover motor branch circuits. Article 440 is listed to cover air-conditioning and refrigeration equipment branch circuits.

ARTICLE 370

OUTLET, DEVICE, PULL AND JUNCTION BOXES, CONDUIT BODIES, AND FITTINGS

370-16 Number of Conductors (b) (1) Box conductor Fill

Revise the wording of the exception. Revised wording: *An equipment grounding conductor or not more than four fixture wires smaller than No. 14, or both, are permitted to be omitted from the calculations when they enter a box from a domed fixture or similar canopy and terminate within the box.*

Reason

The revision makes clear that where six fixture wires smaller than No. 14 enter the box, the calculation would include two fixture wires. When there are more than four fixture wires in this installation, the first four fixture wires are not counted.

370-16(c) Conduit Bodies

Revise the wording and divide into two subdivision.
(1) **General** The raceway fill of a conduit body is limited by the raceway fill permitted for the size of the raceway connected to the conduit body as permitted by Table 1 Chapter 9.
(2) **Splices and Taps** The rewording makes clear that before conduit bodies can contain splices, taps, or devices the *conduit body must be durably and legibly marked by the manufacturer with its cubic-inch capacity. The conduit body fill is calculated according to Section 370-16(b).*

The revision is a clarification.

370-17(c) Nonmetallic Boxes Exception

Where nonmetalic-sheathed cable *or underground branch-circuit and feeder cable* is used with *single-gang* boxes no larger than 2¹/₄ in. by 4 in. mounted in walls or ceilings, the installation is permitted to be made without securing the cable to the box.

Reason

The revision adds the use of underground branch-circuit cable and emphasizes single-gang boxes

370-25(a) Nonmetallic Metal Covers or Plates

A phrase is deleted. Nonmetallic or metal covers and plates shall be permitted. ~~with nonmetallic boxes~~

Result

Nonmetallic or metallic covers are permitted on nonmetallic boxes and metallic boxes. Wherever metal covers are used, they are required to be grounded.

370-27(b) Floor Outlet Boxes Exception

The basic rule requires boxes "listed for floor outlet boxes." Revise the exception by adding:
   Boxes located in elevated floors of show windows and similar locations are permitted to be other than those listed for floor application. Receptacles and covers are required to be listed as an assembly for this type of location.

Reason

The first part of the revision clarifies that there are receptacles with covers designed for this type of installation; and they should be used for it.

370-27 Outlet Boxes

Revise title of subsection (c). 1996 title: **(c) Boxes at Fan Outlets;** 1999 title: **(c) Boxes at Ceiling-Suspended (Paddle) Fan Outlet.** The term *ceiling-suspended (paddle) fan* is also used in the text.

Reason

Paddle fans and exhaust fans can be installed in the same room. This new designation makes clear that this section only applies to paddle fans.

370-70 Boxes Over 600 Volts, Nominal (a) General

Part E of Article 370 covers over 600 volts, nominal. The revision lists the section numbers that apply to box sizing used for conductors under 600 volts, nominal.

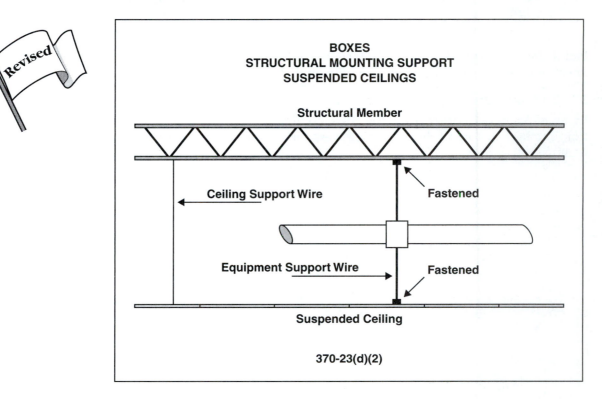

REVISED

**BOXES
STRUCTURAL MOUNTING SUPPORT
SUSPENDED CEILINGS**

Structural Member

Ceiling Support Wire   Fastened

Equipment Support Wire   ➡ Fastened

Suspended Ceiling

370-23(d)(2)

REVISION

The entire Section 370-23, Supports, is revised with rewording, rearrangement, and addition of new information.

Change the words "securely fastened in place" to **"*supported*"** throughout the text. The word "~~adequate,~~" has been deleted.

(a) Surface Mounted Delete ~~Enclosures shall be fastened to the surface upon which they are mounted~~. Replace with ***An enclosure mounted on a building or other surface shall be rigidly and securely fastened in place.***

Delete the title ~~(c) Non-Structural Mounting~~ and replace title and subdivide section.

(b) Structural Mounted
  (1) Nails
  (2) Braces
(c) Mounting in Finished Surfaces
(d) Suspended Ceilings
  (1) Framing Members
  (2) Support Wires
    The support wires must comply with section 300-11(a) for ceiling support wires. The wires must be fastened at both ends and be taut.
      a. Fire-rated assemblies
      b. Non fire-rated assemblies

Reason

This revision brings the *Code* up to date with present-day materials and methods of construction. The subsection on suspended ceilings coordinates with the requirements in *Section 300-11*.

370-23(f) Raceway Supported Fixtures

Exception No. 2 for cantilevered fixtures has one added requirement; *b. The unbroken conduit length before the last point of conduit support is 12 inches or greater, and the portion of the conduit to securely fastened at some point not less than 12 inches from the last point of support.*

370-23(g) Enclosures in Concrete or Masonry

Boxes embedded in concrete are required to have suitable protection from corrosion.

Reason

The corrosion protection for boxes embedded in concrete is required because aluminum embedded in concrete corrodes.

BOXES AT LIGHTING FIXTURE OUTLET   *NEW*

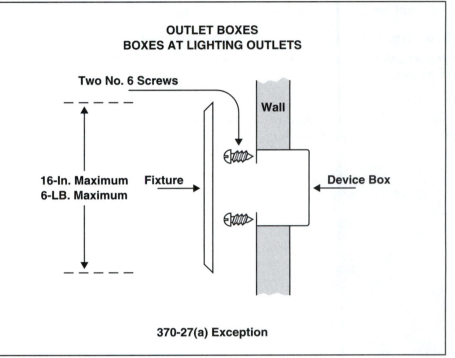

OUTLET BOXES
BOXES AT LIGHTING OUTLETS

Two No. 6 Screws

Wall

16-In. Maximum
6-LB. Maximum   Fixture

Device Box

370-27(a) Exception

BASIC RULE

Boxes used at lighting fixture outlets shall be designed for the purpose. At every outlet used exclusively for lighting, the box is required to be designed for the attachment of a lighting fixture.

NEW EXCEPTION

A fixture weighing not more than 6 lbs. and not exceeding 16 in. in any dimension is now permitted to be mounted on other than a box designed for lighting fixtures, provided it is a wall mounted fixture with not less than two No. 6 screws.

Reason

This revision covers the practice of mounting a small fixture on a device box with two No. 6 screws. A device box is not designed as a lighting outlet; therefore, the exception is needed.

MANHOLES—TUNNELS—ENCLOSURES   370 PART D.

FOR PERSONNEL ENTRY   *NEW*

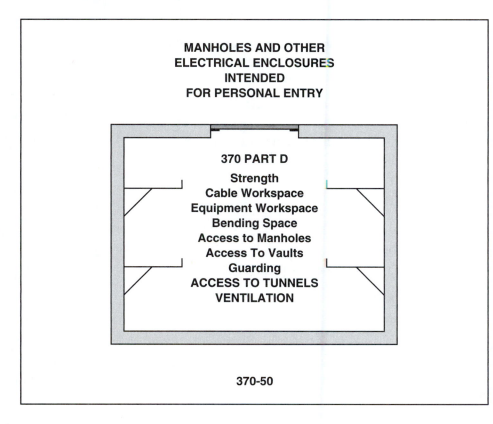

MANHOLES AND OTHER
ELECTRICAL ENCLOSURES
INTENDED
FOR PERSONAL ENTRY

370 PART D

Strength
Cable Workspace
Equipment Workspace
Bending Space
Access to Manholes
Access To Vaults
Guarding
ACCESS TO TUNNELS
VENTILATION

370-50

370-50 General

Sufficient size and work space.
   Exception: Industrial with AHJ approval.

370-51 Strength

Designed under supervised engineering.
   FPN Reference National Electrical Safety Code

370-52 Cabling Workspace

$3^1/2$ ft cables both sides $2^1/2$ ft cables only on one side
   Exception: Communication cable, fire alarm cable, optical fiber cable.

370-53 Equipment Workspace

As in *Sections 310-16* and *310-34.*

370-54 Bending Space for Conductors

600 volts and below as in *Section 370-28(a).*
Over 600 volts as in *Section 370-71(b).*
   Exception: box calculation reference

370-55 Access to Manholes

(a) **Dimensions**
   Rectangular openings 26 in. by 22 in. minimum.
   Round openings 26 in. diameter.
   Exception Opening for special manholes.

(b) Obstructions

Free of protrusions.

(c) Location

Opening not directly above cable or equipment.

(d) Covers

100 lbs. or restrained so they cannot fall in to the manhole.

(e) Markings

Manhole covers marked for function.

370-56 Access to Vaults and Tunnels

(a) Location

Opening not directly above cable or equipment.

(b) Locks

As in Section 110-34(c)

370-57 Ventilation

As needed.

370-58 Guarding

As in Sections 110-27(a)(2) and 110-31(a)(1).

370-59 Fixed Ladder

Corrosion resistant.

Reason

With the deregulation of the utility companies, more and more manholes, tunnels, and enclosures are being used on large premises wiring, and they need to be covered by the *Code*.

Comment

The title of this part of Article 370 has been added to the scope of Article 370. This new part of Article 370 is made up of parts from the National Electrical Code® and the National Electrical Safety Code®. Where needed, dual sectional referencing covers installations both under and over 600 volts, nominal.

CABLES ENTERING ENCLOSURES 373-5(c) Exception

NONMETALLIC-SHEATHED CABLE *NEW*

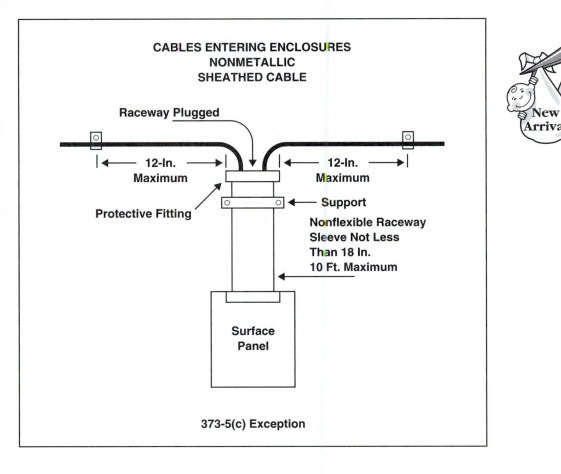

BASIC RULE

Where a cable enters an enclosure, each individual cable is required to be secured to the enclosure.

NEW EXCEPTION

Nonmetallic-sheathed cables are permitted to enter a panel via a conduit sleeve, provided all of the following requirements are met:
1. *The sheath of the cable is entirely nonmetallic.*
2. *The sheath is continuous in the raceway.*
3. *Enclosure is surface mounted.*
4. *One or more nonflexible raceways are permitted.*
5. *The raceway is a minimum of 18 in. in length.*
6. *The maximum length is 10 ft.*
7. *The raceway is plugged at the open end.*
8. *The raceway does not penetrate a ceiling.*
9. *The cables are protected where they enter the raceway.*
10. *The raceway is supported at its outer end.*
11. *The cable is supported within 12 in. of where it enters the raceway. The measurement is along the sheath of the cable.*
12. *Cable fill as per conduit or tubing fill Table 1, Chapter 9.*

Reason

This exception provides a reasonable and safe way for a multitude of non-metallic sheathed cables to enter a panel.

Comment

This new exception applies to a dwelling unit wired with nonmetallic-sheathed cable (Romex) with a multitude of branch circuits coming into the panel. In this installation, one key point is to be sure the raceway into the panel is plugged so nothing can fall down inside the panel. The plugging material to be used is not identified.

ARTICLE 374

AUXILIARY GUTTERS

374-5 Number of Conductors

Delete three exceptions. Reword, outline, and include exceptions in text without changing the intent of the section.

Cross Check

1996	1999
374-5 Number of Conductors	**374-5 Number of Conductors** As per 1 though 4 below.
(a) Sheet Metal Auxiliary Gutters Part of text and *Exception No. 2* Part of text *Exception No. 3* **(b) Nonmetallic Auxiliary Gutters**	**(a) Sheet Metal Auxiliary Gutters** **(1)** **(2)** **(3)** **(4)** **(b) Nonmetallic Auxiliary Gutters**

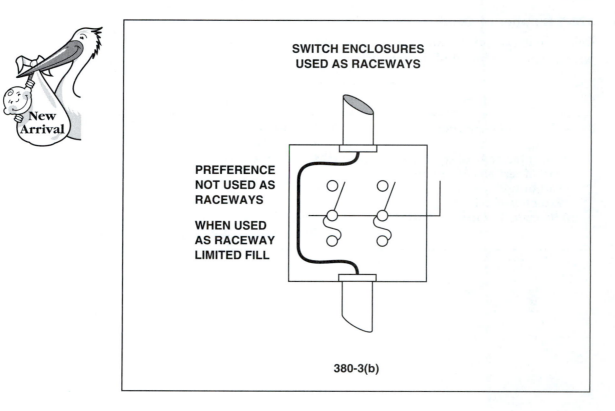

SWITCH ENCLOSURES
USED AS RACEWAYS

PREFERENCE
NOT USED AS
RACEWAYS

WHEN USED
AS RACEWAY
LIMITED FILL

380-3(b)

NEW RULE

380-3(b) Used as a Raceway. Switch enclosures are not permitted to be used as junction boxes or raceways for splicing or pulling through conductors going to other overcurrent devices unless the switch enclosure complies with Section 373-8.

Reason

This revision makes clear that switch enclosures are governed by the same regulations as cabinets and cut-out enclosures.

Comment

Section 373-8 limits the fill of the gutter area of a switch enclosure to a 40% fill for conductors and 75% fill when splices are made.

SNAP SWITCH FACEPLATES *REVISED*

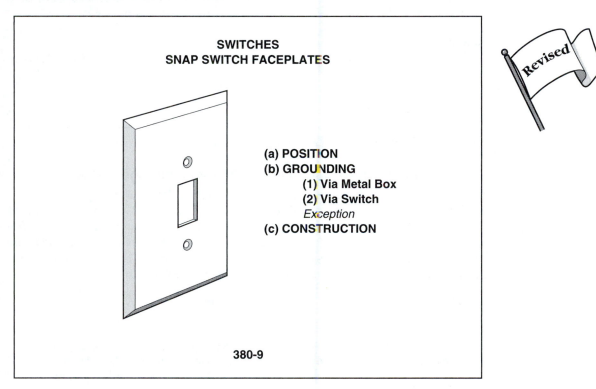

SWITCHES
SNAP SWITCH FACEPLATES

(a) POSITION
(b) GROUNDING
    (1) Via Metal Box
    (2) Via Switch
    Exception
(c) CONSTRUCTION

380-9

REVISION

A long paragraph is broken up, given a new title, outlined, and reworded with additional information.

380-9 Provisions for Snap Switch Faceplates. The faceplates for dimmers are added. When provided by the installation, a metal faceplate is permitted to be grounded via metal screws to a metal box, or the equipment-grounding terminal on the switch or dimmer.

Exception: Where a grounding means does not exist, a snap switch with a grounding means is permitted for replacement only. If the replacement has a metal cover within reach of a conducting floor or surface, a non-metallic noncombustible faceplate is required.

Cross Check

1996	1999
380-9 Faceplates for Flush-Mounted Snap Switches	380-9 Provisions for Snap Switch Faceplates (title only)
	(a) Position (cover wall opening)
Part of text	(b) Grounding Faceplates, snap switches and
Revised text	
New	dimmer switches
Method of grounding	to be grounded
Method of grounding	(1) Via metal box ground terminal
New	(2) Via switch ground terminal
Part of text	Exception (when no ground exists)
	(c) Construction (no change)

380-14(d) AC Specific-Use Snap Switches Rated for 347 Volts

Add a new paragraph to this section: *Specific-use snap switches rated 347 volts are required to be rated at not less than 15 amperes and are not permitted to be interchangeable with general-use snap switches.*

Reason

There was no ampere rating for the 347-volt snap switches in the previous *Code*.

ARTICLE 384

SWITCHBOARDS AND PANELBOARDS

384-3 Support and Arrangement of Busbars and Conductors

(a) Conductors and Busbars on a Switchboard or Panelboard

This section is revised, outlined, and rearranged into an outline without changing the intent of the rule. A new rule is listed under *Service Switchboards: No uninsulated, ungrounded service busbar or service terminal is permitted to be exposed to inadvertent contact by persons or maintenance equipment while servicing the load terminations.*

Cross Check

1996	1999

384-3(a) Conductors and Busbars on a Switchboard or Panelboard
    New
    Part of text
    Part of text and part new
    Part of text
    Exception no change
	384-3(a) Conductors and Busbars on a Switchboard or Panelboard
	    As per (1), (2), and (3) below:
	    (1) Location
	    (2) Service Switchboards
	    (3) Same Vertical Section
	    Exception

PANELBOARDS 384-14(a)

DEFINITION
LIGHTING AND APPLIANCE BRANCH CIRCUIT *NEW*

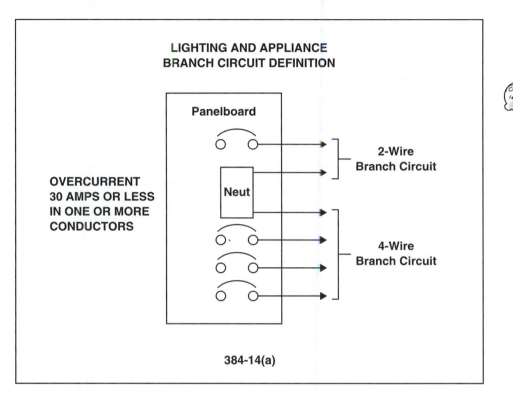

LIGHTING AND APPLIANCE
BRANCH CIRCUIT DEFINITION

Panelboard

OVERCURRENT
30 AMPS OR LESS
IN ONE OR MORE
CONDUCTORS

Neut

2-Wire
Branch Circuit

4-Wire
Branch Circuit

384-14(a)

NEW DEFINITION

A lighting and appliance branch circuit is a branch circuit that has a connection to the neutral of the panelboard and that has overcurrent protection of 30 amperes or less in one or more conductors.

Reason

This definition clarifies that a lighting and appliance branch circuit panelboard may have an unused neutral connection as well as neutral connections that are used. Note that the definition indicates the branch circuit is connected to the neutral in the panelboard.

New
Arrival

```
                    PANELBOARDS
                    DEFINITIONS

   ┌────────────────────────┐   ┌────────────────────────┐
   │   LIGHTING AND          │   │      POWER             │
   │   APPLIANCE             │   │   PANELBOARD           │
   │   BRANCH-CIRCUIT        │   │                        │
   │   PANELBOARD            │   │                        │
   │                         │   │                        │
   │   10% or More           │   │   Less Than 10%        │
   │   Lighting And          │   │   Lighting And         │
   │   Appliance             │   │   Appliance            │
   │   Branch Circuits       │   │   Branch Circuits      │
   └────────────────────────┘   └────────────────────────┘

                       384-14(b)
```

NEW DEFINITION

(b) Power Panelboard: A power panelboard is one having 10 percent or fewer of its overcurrent devices protecting lighting and appliance branch circuits.

REVISED DEFINITION

(a) Lighting and Appliance Branch Circuit Panelboard: A lighting and appliance branch-circuit panelboard has more than 10% of its overcurrent devices protecting lighting and appliance branch circuits.

Reason

This new definition clarifies the distinction between a lighting and appliance branch circuit panelboard and a power panelboard.

Comment

The new definition of a "lighting and appliance branch circuit" should clarify the use of a lighting and appliance branch-circuit panelboard, and a power panelboard. Even though a lighting and appliance branch circuit panel board has a neutral bar in it, a power circuit can originate in it. When a power panelboard has a neutral bar in it, a lighting and appliance branch circuit can originate in the power panelboard.

POWER PANELBOARD

OVERCURRENT PROTECTION

384-16(b)

NEW

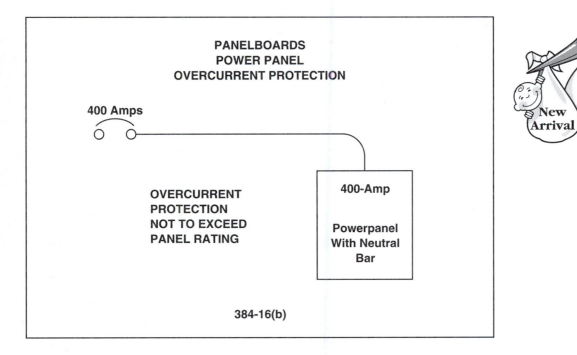

**PANELBOARDS
POWER PANEL
OVERCURRENT PROTECTION**

400 Amps

**OVERCURRENT
PROTECTION
NOT TO EXCEED
PANEL RATING**

400-Amp

**Powerpanel
With Neutral
Bar**

384-16(b)

New
Arrival

NEW RULE

384-16(b) Power Panelboard Protection. In addition to a panelboard having a rating of not less than the load calculated according to Article 220, a power panelboard with neutral connections available and more than 10% of its overcurrent devices protecting branch circuits rated 30 amperes or less shall be protected on the supply side by an overcurrent protective device having a rating not greater than that of the panelboard.

NEW EXCEPTION

384-16(b) Exception: This individual protection shall not be required for a power panelboard used as service equipment with multiple disconnecting means in accordance with Section 230-71.

Note

With the insertion of a new Section 384-16(b) the existing (b), (c), (d), (e), and (f) are now (c), (d), (e), (f), and (g), respectively.

Reason

Overcurrent should be addressed with panelboards. The concern with less than 30-ampere branch circuits is based on Sections 430-62(a) and 430-53. The exception is made to cover power panels used for service equipment.

Comment

Although there is a neutral bar in the power panel, it could have branch circuits rated less than 30 amperes that are not lighting and appliance branch circuits. *Sections 430-62(a)* and *430-63* concern motor feeder ground-fault and short-circuit protection.

   Section 230-71 permits six disconnects to be used as the service disconnecting means.

CHAPTER 4
EQUIPMENT FOR GENERAL USE

■ ■ ■ ■ ■ ■ ■ ■ ■ ■ ■ ■ ■ ■ ■ ■ ■ ■ ■

ARTICLE 400

FLEXIBLE CORDS AND CABLES

Table 400-4

Add *a new portable Power Cable Type G-GC*. To introduce this type of cable, several change are made in *Article 400* to recognize and set standards for it.

Comment

G-GC cable consists of three conductors plus two grounding conductors and one ground check conductor. The ground check conductor monitors the continuity of the grounding conductor and is permitted to be not smaller than No. 10 AWG. The introduction of this new cable creates three new exceptions for wire size, for insulation thickness, and for over 600 volts, nominal use.

400-7 Uses Permitted

New use: (3) Flexible cord is now permitted for the connection of portable lamps, *portable and mobile signs,* or appliances.

400-8 Uses Not Permitted

List new restrictions: (2) Flexible cord is not permitted where run through holes in walls, *structural* ceilings, *suspended ceilings, dropped ceilings*, or floor.

Reason

This revision makes clear that flexible cords or cables are not to be run through any type of ceiling.

ARTICLE 402

FIXTURE WIRES

Table 402-3 Fixture Wires

This revision replaces ~~latex rubber~~ with the term ***thermoset*** and ~~Heat-Resistant Latex Rubber~~ with ***Cross-Linked Synthetic Polymer.***

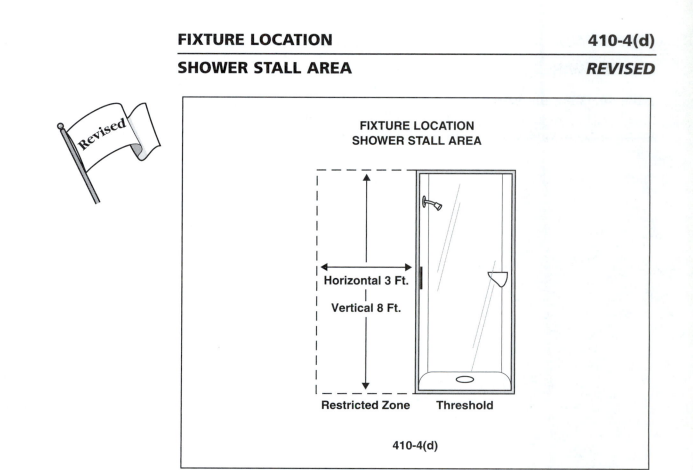

**FIXTURE LOCATION
SHOWER STALL AREA**

Horizontal 3 Ft.

Vertical 8 Ft.

Restricted Zone   Threshold

410-4(d)

REVISION

(d) Above Bathtubs *and Shower Areas.* No part of cord-connected fixtures, hanging fixtures, lighting track, pendants, or ***ceiling-suspended (paddle)*** fans shall be located within a zone measured 3 ft horizontally and 8 ft vertically from the top of the bathtub rim ***or shower stall threshold***. This zone is all encompassing, and includes the zone directly over the tub ***or shower stall***.

DELETION

The last two paragraphs of Section 410-4(a) are deleted. The paragraphs defined damp and wet locations and that should be located in Article 100 Definitions.

Reason

The revision extends the coverage to the shower stall because it presents just as much of an electrical hazard as the bathtub. The term "paddle" is inserted to make clear that the application is for ceiling-suspended paddle fans not recessed exhaust fans.

LIGHTING FIXTURES   410-11

BRANCH CIRCUITS THROUGH JUNCTION BOX   *REVISED*

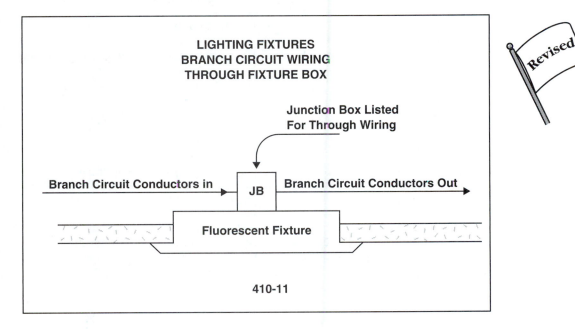

**LIGHTING FIXTURES
BRANCH CIRCUIT WIRING
THROUGH FIXTURE BOX**

Junction Box Listed
For Through Wiring

Branch Circuit Conductors in   JB   Branch Circuit Conductors Out

Fluorescent Fixture

410-11

REVISION

Branch-circuit wiring, *other than 2-wire or multiwire branch circuits supplying power to fixtures connected together*, shall not be passed through an outlet box that is an integral part of ~~an incandescent~~ fixture unless the fixture is identified for through-wiring. Add a new FPN: ***See Section 410-31 Exceptions for circuits supplying power to fixtures connected together.***

Reason

Electrical discharge type lighting fixtures, such as fluorescent fixtures, are now made with a box as an integral part of the fixture and are listed for pass through branch-circuit wiring.

Comment

The identification of the 2-wire and multiwire branch circuits used in the revision is the same as for fixtures designed to be assembled end-to-end. As the FPN indicates, it is located in Section 410-31.

410-14 Connection of Electrical-Discharge Lighting Fixtures

(a) Independent of Outlet Box The exception permitting flexible cord connections is deleted and made a part of the text without a change in the intent of the rule.

410-15 Supports

Section 410-15(b) Supports for Metal Poles Supporting Lighting had a number of subdivisions and exceptions with a multitude of separate parts. The revision breaks this section into an outline for ease of interpretation. There is no change in the intent of the rules.

Cross Check

1996	1999
D. Fixture Supports	**D. Fixture Supports**
410-15 Supports (a) **General** (b) **Metal Poles Supporting Lighting Fixtures** **(1)** **(2)** *Exception* **(1)** *Exception*	**410-15 Supports** (a) **General** (b) **Metal Poles Supporting Lighting Fixtures** **(1)** Part of text *Exception No. 1* *Exception No. 2*
(1) **(3)** Ground terminal **(2)** Access handhole **(1)** Hinged base **(2)** *Exception*	**(2)** Part of text **(3)** (a) Access handhole (b) Hinged base *Exception*
(1) *Exception* **(3)** Grounding as per 250 **(4)** Conductors in vertical poles	**(4)** Bond-hinged pole **(5)** Grounding as per 250 **(6)** Conductors in vertical poles

LIGHTING POLES   410-15(b)

METAL POLES ARE RACEWAYS   *REVISED*

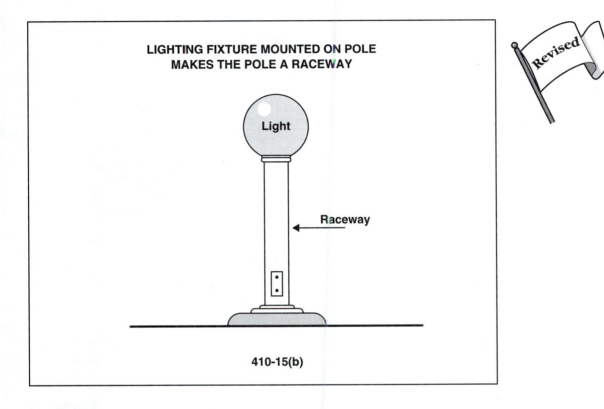

LIGHTING FIXTURE MOUNTED ON POLE
MAKES THE POLE A RACEWAY

Light

← Raceway

410-15(b)

REVISION

410-15 Metal Poles Supporting Lighting Fixtures Metal poles are permitted to be used to support lighting fixtures and *as a raceway to* enclose supply conductors as follows:

Reason

Opinions differed about whether a metal pole was a raceway or not. This revision clarifies that the metal pole supporting a fixture is indeed a raceway, and the conductors are enclosed in a raceway.

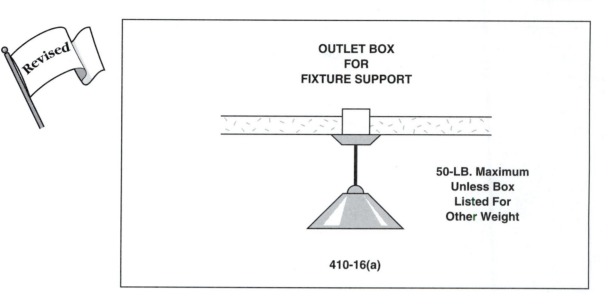

OUTLET BOX
FOR
FIXTURE SUPPORT

50-LB. Maximum
Unless Box
Listed For
Other Weight

410-16(a)

REVISION

A fixture weighing more than 50 lbs. is required to be independent of the outlet box **unless the outlet box is listed for the weight to be supported.**

Reason

Some outlet boxes are designed and listed for the safe support of fixtures over 50 lbs.

410-16(b) Inspection

Delete the exception and make it a part of the text. Lighting fixtures are required to be installed so that they can be inspected without disconnecting the fixture, **unless the fixtures are connected by attachment plugs and receptacles.**

410-16(c) Suspended Ceilings

The clips identified for use with the type of ceiling framing members and fixtures are now required to be **listed**.

Reason
Safety.

410-18 Grounding Exposed Fixture Parts

(a) Exposed Conductive Parts ~~Exposed Conductive parts of lighting fixtures and equipment directly wired or attached to outlets supplied by a wiring method that provides an equipment ground shall be grounded.~~

Revised wording: **Exposed metal parts shall be grounded or insulated from ground or other conducting surfaces or inaccessible to unqualified personnel. Lamp tie wire, mounting screws, clips, and decorative bands on glass spaced at least 1 1/2 in. (38 mm) from lamp terminals are not required to be grounded.**

Comment

The first sentence is reworded for clarification. The second sentence, "Lamp ties wire . . . ." was an exception following Section 410-19. **Section *410-19*, Equipment Over 150 Volts to Ground,** is completely deleted.

Reason

Section 410-19 is deleted because it was unenforceable.

410-35 Fixture Rating

(a) Marking The wording and temperature ratings are revised. All fixtures ~~requiring ballast or transformers~~ are required to be marked with ***the maximum lamp wattage***, or their electrical rating and the manufacturer's name, trademark, or other suitable means of identification. A fixture requiring supply wire rated higher than ~~90°C (194°F)~~ 60°C (140°F) shall be so marked in letters not smaller than 1/4 in. high on equipment or carton.

Reason

By deleting ballast or transformers, this revised rule now applies to all lighting fixtures. This revision of the temperature rating derives from concern about the fire hazard of using the wrong temperature rating on the wire connecting fixtures. Many incandescent lighting fixtures require 75° C or 90° C conductors. The previous rule only required the marking of fixtures rated higher than 90°C.

LIGHTING FIXTURES

410-42(b)(5)

GROUNDING PORTABLE HANDLAMPS

NEW

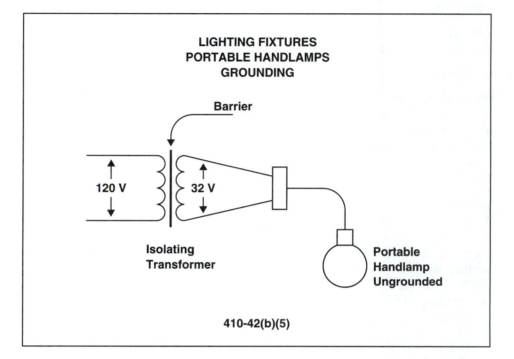

LIGHTING FIXTURES
PORTABLE HANDLAMPS
GROUNDING

Barrier

120 V

32 V

Isolating
Transformer

Portable
Handlamp
Ungrounded

410-42(b)(5)

NEW RULE

Portable handlamps are not required to be grounded where supplied through an isolating transformer with an ungrounded secondary of not over 50 volts.

Reason

This revision coordinates with Section 250-114(4)(g) Exception, which also permits this installation.

Comment

This type of installation is used in wet or damp locations and for working on the inside of large tanks or vessels. This handlamp would also have a guard on it.

DELETION

410-51 Lead Wires This section and the exception are deleted as they do not apply to *Code* construction requirements. They are manufacturing requirements, covered under product standards.

RECEPTACLES   410-56

TYPES AND INSTALLATION   *REVISED*

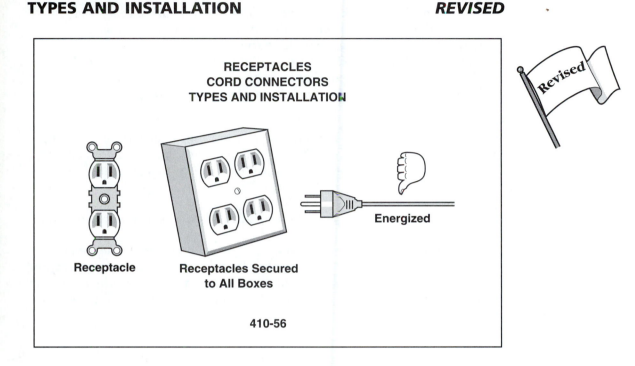

RECEPTACLES
CORD CONNECTORS
TYPES AND INSTALLATION

Receptacle

Receptacles Secured
to All Boxes

Energized

410-56

REVISIONS

410-56 (a) Receptacle. Delete the exception permitting the use of a 10-ampere, 250-volt receptacle.

410-56(c) Isolated Ground Receptacles Isolated ground receptacles installed in nonmetallic boxes are required to be covered *either* with a nonmetallic faceplate. An exception permits the use of *an effectively grounded metal faceplate.*

410-56 (g) **Attachment Plugs** Add a new sentence: *Attachment plugs are required to be so installed that their prongs, blades, or pins are not energized unless inserted into the required energized receptacle. No receptacle is permitted to be installed so as to require an energized attachment plug as its source of supply.*

Reason

The 10-ampere, 250-volt receptacle is deleted as they are no longer manufactured. Energized attachment plug caps have occurred in recreation vehicles and other power supply cords. They are very hazardous.

410-56(f)3 Receptacles ~~in Raised~~ **Mounted** on **Covers** ~~Receptacles installed in raised covers shall not be secured solely by a single screw.~~ *Receptacles mounted to and supported by a cover are required to be secured by more than one screw or shall be a device assembly or box cover listed and identified for securing by a single screw.*

REVISION

410-57(b) Wet Locations

(1) Cord and plug connection inserted into a wet location receptacle is required to be weatherproof if operating is unattended, such as sprinkler, landscaping, or holiday lighting.

(2) Cord and plug connection inserted into a wet location receptacle and the equipment is intended to be used attended is not required to be weatherproof. Example of use is an electric lawn edger.

Reason

This revision addresses the concern for cord and plug connected equipment in a wet location left unattended. This revision clarifies the use of cord and plug connected equipment being used in a wet location, indoors or outdoors.

NEW RULE

410-58 Grounding-Type Receptacles, Adapters, Cord Connectors, and Attachment Plugs

   (a) Grounding-Pole Identification. Add a new sentence: *The grounding conducting pole of grounding-type plug in ground-fault circuit-interrupter is permitted to be of the movable, self-restoring type on circuits operating at not over 150 volts between conductors nor over 150 volts between any conductor and ground.*

Reason

This type GFCI is now on the market. It can be plugged into an existing non-grounded, 2-wire receptacle to give ground-fault protection for personnel.

Comment

This type installation gives ground-fault protection for personnel but not for equipment, because it lacks an equipment grounding conductor for the return path to the overcurrent device protecting the equipment.

410-66 Clearance and Installation

(a) Clearance. The two exceptions are reworded into permissive rules and subdivided as part of the text without changing the intent of the rule.

   (b) Installation. The exception is reworded into a permissive rule and made a part of the text without a change in the intent of the rule.

FIXTURES   410-67(c)

TAP CONDUCTORS   *REVISED*

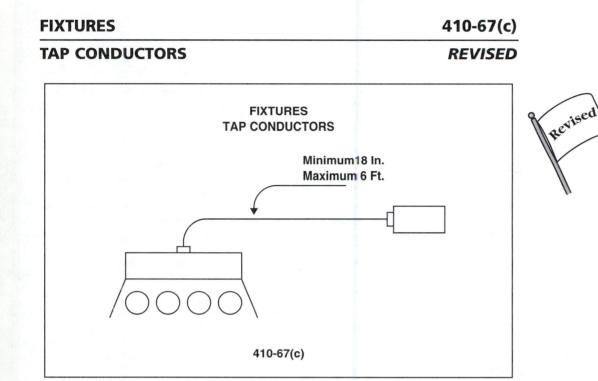

FIXTURES
TAP CONDUCTORS

Minimum 18 In.
Maximum 6 Ft.

410-67(c)

REVISION

(c) Tap Conductors Revision: Such tap **conductors** shall be in a suitable raceway or Type AC or MC cable of at least ~~4 ft~~ *18 in. (450 mm)* but not more than 6 ft.

Reason

The outlet box is required to be at least 1 ft from the fixture. With an outlet box located 1 ft from the fixture, a minimum of 4 ft of tap conductor can cause the accumulation of cable in the ceiling as well as other heat-related problems.

Comment

This is a big change in a long-standing rule. The 4-ft tap was an asset for fixture maintenance.

~~410-69 Housing.~~ Delete this entire section and its exception. The rule determined metal thickness used for fixture housings, but it was unenforceable because it related to product standards.

410-73(e) Thermal Protection Delete the three exceptions and rearrange into an outline. Put the exceptions in the text without changing the intent.

Cross Check

1996	1999
410-73(e) Thermal Protection (basic rule) only)	**410-73(e) Thermal Protection** (title
	(1) (basic rule)
Exception No. 1	**(2)**
Exception No. 2	**(3)**
Exception No. 3	**(4)**
410-73(f) High-Intensity Discharge Fixtures (basic text)	**410-73(f) High-Intensity Discharge Fixtures** (title only)
	(1) Part of basic text
(FPN)	**(2)**
Exception	**(3)**
	(4) Part of basic text

ARTICLE 422

APPLIANCES

Article 422, Appliances, is rearranged with very few section numbers remaining the same. Many of the exceptions are made into permissive rules and added to the text. The new titles for the various parts indicates the system used for revision.

Cross Check

1999	1999
Part A. General	Part A. General
Part B. Branch-Circuit Requirements Part C. Installation of Appliances	Part B. Installation
Part D. Control and Protection of Appliances	Part C. Disconnecting Means
	Part D. Construction
Part E. Marking of Appliances	Part E. Marking

Reason

The revision rearranges the text in a more logical order to make it more user-friendly without changing the intent of the rules.

NEW RULE

422-15 Central Vacuum Outlet Assemblies
(a) Listed central vacuum outlet assemblies are permitted to be connected to a branch circuit in accordance with Section 210-23(a).
(b) The ampacity of the connecting conductors is the same as the ampacity of the branch circuit conductors.
(c) Any exposed noncurrent-carrying metal parts of the assembly are required to be grounded.

Reason

This new rule recognizes a listed product currently being installed.

Comment

Section 210-23(a) defines limits for permissible loads on 15- and 20-ampere circuits. It limits any plug- and cord-connected piece of equipment to 80 percent of the branch circuit rating. The unusual feature of this outlet is that the branch circuit is completed within the wall once the vacuum has been plugged into the outlet.

DISHWASHER

CORD LENGTH

422-16(b)(2)(b)

REVISED

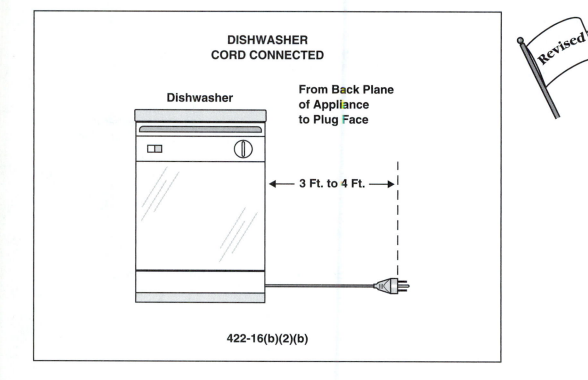

DISHWASHER CORD CONNECTED

Dishwasher

From Back Plane of Appliance to Plug Face

◄— 3 Ft. to 4 Ft. —►

422-16(b)(2)(b)

REVISION

The length of the cord for a cord connected dishwasher is required to be 3 ft to 4 ft in length *measured from the face of the attachment plug to the plane of the rear of the appliance.*

Reason

This measurement is in accord with product standards. For field installation, the connection on the dishwasher may be on the opposite side from a receptacle located at an accessible location under the sink.

CEILING-SUSPENDED (PADDLE) FANS

422-18

LISTED OR NONLISTED

REVISED

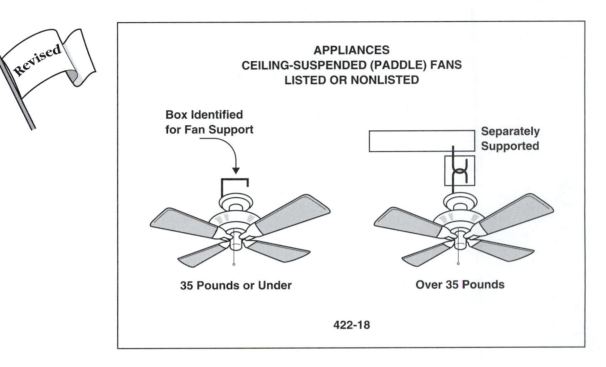

REVISION

422-18 Support of Ceiling-*Suspended (Paddle) Fans.*
(a) Ceiling-*Suspended (Paddle)* Fans 35 Lb. (15.88 kg) or Less Delete the word "Listed" and revise wording so as to identify fans as "ceiling-suspended." ~~Listed~~ Ceiling-***suspended (paddle)*** fans that do not exceed 35 lbs. etc.
(b) Ceiling-*Suspended (Paddle)* Fans Exceeding 35 lb. (15.88 kg)

Reason
The term "ceiling-suspended fan" more clearly identifies the type of fan the requirement applies to. The deletion of the word "Listed" now makes this section apply to the mounting of both listed and nonlisted ceiling-suspended fans. The term "paddle" is inserted in parenthesis to better identify with field terminology.

Comment
The 35 lb. limitation includes any lighting kit or fixture that might be mounted below the fan.

NEW EXCEPTION
Exception: Listed outlet boxes or outlet box systems that are identified for the purpose shall be permitted to support listed ceiling fans with or without accessories, weighing not more than 70 lb. (31.76 kg).

Reason
Some listed outlet boxes have been tested for this purpose.

ARTICLE 424

FIXED ELECTRICAL SPACE-HEATING EQUIPMENT

All the exceptions in Article 424 are deleted, reworded, and made a part of the preceding text as positive statements without changing the rules.

The two sections on grounding referencing Article 250 are deleted as they are redundant to the requirements in Article 250.

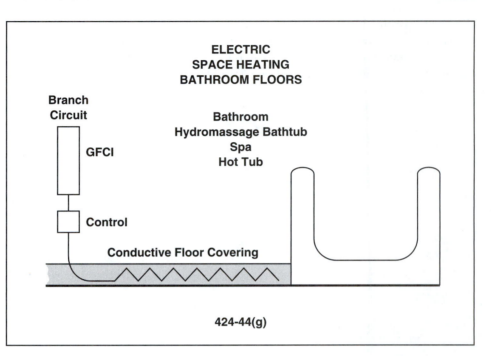

NEW RULE

(g) Ground-Fault Circuit-Interrupter Protection for Conductive Heated Floors of Bathrooms, Hydromassage Bathtubs, Spa, and Hot Tub Locations. GFCI protection is now required in all of the above locations. This rule applies to all systems used with conductive floor coverings, whether cable, panel, or other approved heating means is installed.

Reason

Concrete floors are subject to cracking. Thus they may become energized when wet. In Section 110-26(a)(1) Condition 1, the *Code* considers a concrete surface to be a grounded surface.

ARTICLE 426

FIXED OUTDOOR ELECTRIC DE-ICING AND SNOW-MELTING EQUIPMENT

All the exceptions in Article 246 are deleted, reworded, and made a part of the preceding text as permissive rules without changing the rules.

The sections on grounding referencing Article 250 is deleted as it is redundant to the requirements in Article 250.

426-28 Equipment Protection—REVISED

Ground-fault protection of equipment shall be provided for ~~branch circuits supplying~~ fixed outdoor electric de-icing and snow-melting equipment.

Reason

The ground-fault protection can be accomplished with a receptacle. It need not cover the entire branch circuit supplying the outdoor electric de-icing and snow-melting equipment.

Comment

Section 210-8(a)(3) requires all outdoor receptacles to have ground-fault circuit-interrupter protection and gives an exception for a receptacle used for outdoor electric de-icing and snow-melting equipment.

ARTICLE 427

FIXED ELECTRIC HEATING EQUIPMENT FOR PIPELINES AND VESSELS

427-22 Equipment Protection Revision

Ground-fault protection of equipment is required to be provided for each branch circuit supplying electrical ~~heating equipment~~ *heat tracing and heating panels.*

Reason

The previous wording applied to all resistance heating. The revision more clearly focuses on potential hazards.

427-23 Metal Covering

Relocate a sentence from the end of the section to the beginning so that it applies to both parts **(a) Heating Wires or Cables** and **(b) Heating Panels.**

All the exceptions in Article 427 are deleted, reworded, and made a part of the preceding text as permissive rules without changing the intent of the rules.

The section 427-21 on grounding referencing Article 250 is deleted as it is redundant to the requirements in Article 250.

ARTICLE 430

MOTORS, MOTOR CIRCUITS, AND CONTROLLERS

430-1 Scope

Delete the two exceptions and combine them into one FPN. Reference Section 110-26(f) for installation requirements and Article 440 for air-conditioning and refrigeration equipment.

430-14 Location of Motor

(b) Open Motor Delete as part of text and restate as a new exception. *Exception: Installation of these motors on wood floors or supports shall be permitted.*

Comment

This takes a statement out of the text and makes it into an exception without changing the rule.

Some of the letter identification of the **Parts of Article 430** are changed.

Cross Check

1996	1999
Part I. Disconnecting Means	J. Disconnecting Means
Part J. Over 600 Volts, Nominal	K. Over 600 Volts, Nominal
Part K. Protection of Live Parts All Voltages	L. Protection of Live Parts All Voltages
Part L. Grounding All Voltages	M. Grounding All Voltages
Part M. Tables	N. Tables

Reason

The letters I and O are easily confused with the numbers 1 and 0; therefore, they are deleted.

Exceptions

Many of the exceptions in Article 430 are reworded into complete sentences. Example: Table 430-37 *Exception: Where protected by other approved means.* Rewording: *Exception: An overload unit in each phase shall not be required where overload protection is provided by other approved means.*

Tables

For easier reading, horizontal lines have been added to all the motor ampacity tables.

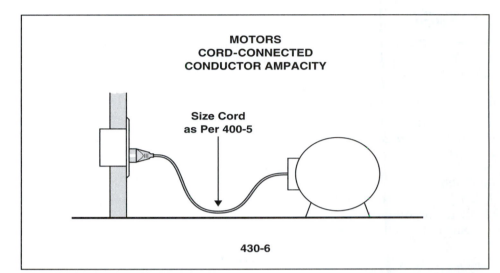

**MOTORS
CORD-CONNECTED
CONDUCTOR AMPACITY**

Size Cord
as Per 400-5

430-6

NEW RULE

430-6 Ampacity and Motor Rating Determinations Add a new sentence to indicate: *"Where flexible cord is used to connect a motor, the ampacity of the conductors is selected according to Section 400-5."*

Reason

Motors are permitted to be cord- and plug-connected. The basic rule indicates the motor conductor ampacity is to be based on Tables 310-16 through 310-19, which do not apply to cords.

Comment

Section 430-6 identifies the current values to be used to determine the ampacity or rating of the motor disconnecting means, the branch-circuit conductors, the controller, the branch- circuit short-circuit and ground-fault protection, and any separate overload protection. Section 400-5 refers to Tables 400-5A and 400-5B for the ampacity of various types of cords.

REVISION

This section is rearranged into an outline form, making it easier locate requirements.

Cross Check

1996	1999
430-6 Ampacity and Motor Rating Determination	**430-6 Ampacity and Motor Rating Determination** (revise and add to)
(a) General Motor Applications (part)	**(a) General Motor Application** (Part) As per (1) and (2) below.
(a) General Motor Applications (part) *Exception No. 1* *Exception No. 2*	**(1) Table Values** (revised) *Exception No. 1* *Exception No. 2* ***Exception No. 3*** (new)
(a) General Motor Applications (part) **(b) Torque Motors** FPN	**(2) Nameplate Values** **(b) Torque Motors** FPN
(c) AC Adjustable Voltage Motor	**(c) AC Adjustable Voltage Motor**

MOTOR CURRENT RATINGS   430-6(a)(1) Exception No. 3

MOTOR-OPERATED APPLIANCES   *NEW*

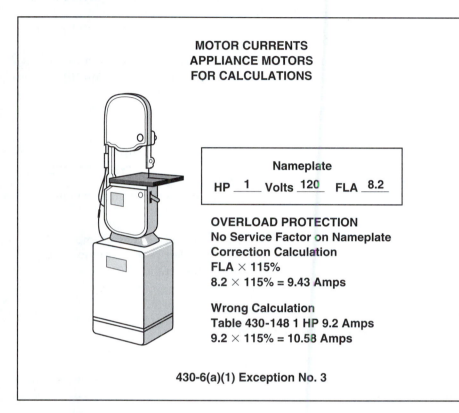

MOTOR CURRENTS
APPLIANCE MOTORS
FOR CALCULATIONS

Nameplate
HP __1__ Volts _120_ FLA _8.2_

OVERLOAD PROTECTION
No Service Factor on Nameplate
Correction Calculation
FLA × 115%
8.2 × 115% = 9.43 Amps

Wrong Calculation
Table 430-148 1 HP 9.2 Amps
9.2 × 115% = 10.58 Amps

430-6(a)(1) Exception No. 3

New Arrival

BASIC RULE

Motor full-load current Tables 430-147, 430-148, 430-149, and 430-150 are to be used for the various motor calculations.

NEW EXCEPTION

When a listed motor-operated appliance is marked with both motor horsepower and full-load current, use the motor full-load current marked on the nameplate of the appliance to determine the ampacity or rating of the disconnecting means, the branch-circuit conductors, the controller, the branch-circuit short-circuit and ground-fault protection, and any separate overload protection.

Reason

Some appliances are marked with a deceptive motor horsepower rating designed to sell the motor but not giving the true horsepower value. Such a horsepower marking might be a peak horsepower marking, rather than its regular horsepower. The nameplate current rating is the actual full-load current rating.

Comment

When a motor is listed, the listing is based on the full-load current, not the nameplate horsepower.

Motor Circuit Conductors

430-22 Single Motor Revision

(a) General Revise wording: Branch-circuit conductors employing a single motor ***used in a continuous duty application*** shall have an ampacity of not less than 125 percent of the full-load current of the motor as determined by Section 430-6(a)(1).

Deletion: With the identification of the "continuous duty application," Exception No. 1 is deleted, reworded, and made a new subsection **(b)** because it was an exception for motors other than continuous duty motors. Exceptions 2 and 3 are now Exceptions 1 and 2, respectively.

Cross Check

1996	1999

430-22 Single Motor
(a) General
  Exception No. 2
  Exception No. 3
 	
  Exception No. 1
	430-22 Single Motor
	(a) General (continuous motors)
	  Exception No. 1
	  Exception No. 2
	  (FPN) New. See Example D8.
	(b) Other Than Continuous Duty
	  (new title)

Reason

This revision and deletion clarifies the rule.

430-40 Overload Relays. Delete Exception No. 2.

~~Exception No. 2. The fuse or circuit breaker ampere rating shall be permitted to be marked on the nameplate of approved equipment in which overload relay is used.~~

Reason

430-40 is directed at overload relay settings and not overload relays.

430-43 Automatic Restart—REVISION

A motor ***overload device*** that can restart a motor automatically after motor ***tripping*** is not permitted to be installed if the automatic restarting of the motor can cause injury.

Reason

This revision clarifies the intent of the rule. Although Section 430-43 is under Part C, Motor Branch-Circuit Overload Protection, it was being wrongly interpreted to mean that any motor that automatically stopped by any means could not be restarted.

430-52 Rating and Setting for Individual Motors.

(Motor Branch-Circuit Short-Circuit and Ground-Fault Protection)

(c) Rating or Setting. (3) *Exception.* The exception now permits ***Design B energy efficient*** motor protection to be the same as Design E motor and not to exceed 1700 percent of the motor's full-load current.

Reason

The Design B energy efficient motor is considered to be just as efficient as the Design E motor and has some of the same characteristics.

430-72 Overcurrent Protection

This section has been rearranged. Seven exceptions are deleted and reworded into permissive rules. Now only two exceptions remain. The intent of the rules is unchanged.

Cross Check

1996	1999
430-72 Overcurrent Protection **(a) General**	**430-72 Overcurrent Protection** **(a) General**
(b) Conductor Properties As per Table 430-72(b). *Exception No. 4* *Exception No. 3* *Exception No. 1* *Exception No. 2*	**(b) Conductor Properties** As per (1) and (2) below *Exception No. 1* (control circuit) *Exception No. 2* (2-wire transformer) **(1)** Table 430-72(b) column A use **(2)** Table 430-72(b) column B use Table 430-72(b) column C use
(c) Control Transformer As per Article 450. *Exception No. 5* *Exception No. 3* **(c) Control Transformer** *Exception No. 1* *Exception No. 2* *Exception No. 4*	**(c) Control Circuit Transformer** As per (1), (2), (3), (4), and (5). *Exception No. 1* (opening hazard) **(1)** Class 1, 2, and 3 circuits **(2)** As per 450-3 **(3)** Transformer 50 VA or less **(4)** 2-amp primary **(5)** Other means

430-83 Motor Controller Ratings

This section has been rearranged and reworded. Five exceptions are deleted and reworded into permissive rules. There is no change in the intent of the rules.

Motor Controllers

Cross Check

1996	1999
430-83 Ratings (title only) **(a) Horsepower Rating at Application** **Voltage** *Exception No. 1* *Exception No. 3*	**430-83 Ratings** As per (a), (b), (c), and (d) **(a) General** (title only) **(1)** **(2)**
	(b) *Small Motors* (new title) *As per 430-81(b).*
Exception No. 2	**(c) Stationary Motors of 2 HP or Less** **(1)** 2 × FLA **(2)** 80% switch ampacity
Exception No. 4 **(b) Voltage Rating**	**(d) Torque Motors** **(e) Voltage Rating**

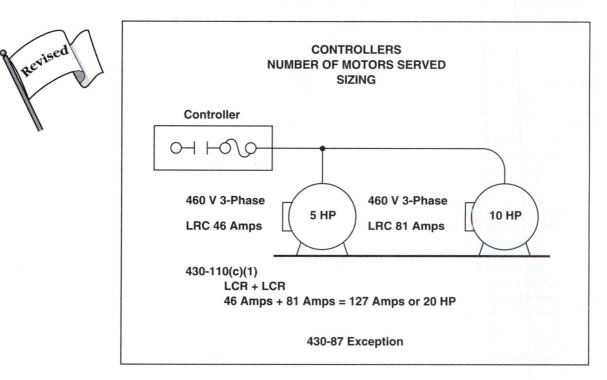

BASIC RULE

430-87 Number of Motors Served by Each Controller Each motor is required to be served by an individual controller.

REVISION

*Exception: For motors rated 600 volts or less, a single controller rated at not less than the equivalent horsepower of all the motors in the group, **as determined in accordance with Section 430-110(c)(1),** is permitted to serve the group under any one of the following conditions:*
a. All the motors are on a single machine.
b. All the motors are protected by one overcurrent protective device.
c. All the motors are in single room in sight from the controller.

Reason

The horsepower rating of a controller is now calculated like that of a motor disconnecting switch.

Comment

When the above calculations are based on Section 430-10(c)(1) the locked rotor current (LRC) is used. This may give a different result than adding the horsepower rating of the motors together.

MOTOR CONTROL CENTERS 430-97(b)

PHASE ARRANGEMENT—HIGH LEG *REVISED*

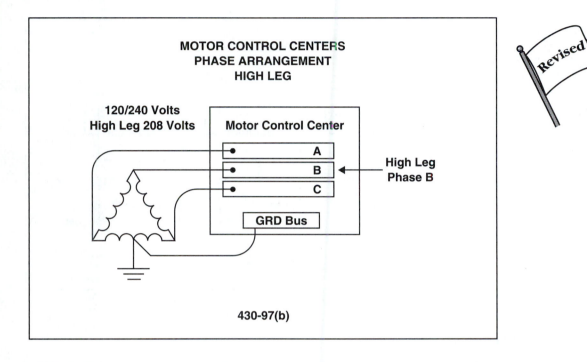

REVISION

The phase arrangement on 3-phase *horizontal common power and vertical buses* is required to be A, B, C, from front to back, top to bottom, or left to right, as viewed from the front of the motor control center. *The high leg is required to be B phase on a 3-phase 4-wire delta connected system. Other bus bar arrangements are permitted for additions to existing installations and shall be marked.*

Reason

The term "common power buses" is inserted to distinguish buses from conductors. The high leg did not previously have to be identified on a motor control center. This revision brings motor control centers into line with switchboards, Section 384-3(f).

SEPARATE DISCONNECTS *REVISED*

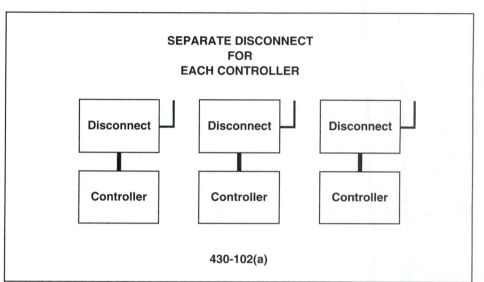

SEPARATE DISCONNECT
FOR
EACH CONTROLLER

Disconnect	Disconnect	Disconnect
Controller	Controller	Controller

430-102(a)

REVISION

~~A disconnecting means was required to be located in sight from the controller location and shall disconnect the controller.~~

An individual disconnecting means is required to be provided for each controller and shall disconnect the controller. The disconnect means is required to be in sight from the controller location.

Reason

The previous rule could be interpreted as allowing a one disconnecting means for a multitude of controllers.

NEW EXCEPTION

430-102(a) Exception No. 2: A single disconnecting means is permitted for a group of coordinated controllers that control several parts of a single piece of equipment.

REVISION
Part J. Disconnecting Means.

430-109 Types. The eight exceptions in this section are deleted. The section is rearranged and reworded into an outline, and exceptions are turned into permissive rules. Two new types of motor disconnecting means are introduced. A new FPN is added giving a descriptive definition of a combination controller.

Cross Check

1996	1999
430-109 Types	**430-109 Types** Reworded Check (a) through (g)

430-109 Types (part)
Exception No. 1
430-109 Types (part)
 	
430-109 Types (part)
Exception No. 8
New
 	
New
	(a) General (2itle only)
	    (1) HP-rated motor circuit switch
	        Design E motors (included in text)
	    (2) Circuit breaker
	    (3) Molded case switch
	    (4) Instantaneous trip CB
	    (5) *Self protected combination*
	    (6) *Manual motor controller marked "Suitable as Motor Disconnect"*

Exception No. 2
 	
Exception No. 3 (Part 1)
Exception No. 3 (Part 2)
New
	(b) Stationary Motors 1/8 HP or less
	(c) Stationary Motors 2 HP or less
	    (1) General use switch 2 HP or Less
	    (2) 80% of snap switch rating
	    (3) HP Rated Controller

Exception No. 4
 	
Exception No. 4a.
Exception No. 4b.
Exception No. 4c.
	(d) Autotransformer-Type Controlled Motors
	    (1) Generator with overloads
	    (2) Controller with limitations
	    (3) Separate fuse or CB

Exception No. 5 *Exception No. 6* *Exception No. 7*	**(e) Isolating Switches** **(f) Cord- and Plug-Connected Motors** **(g) Torque Motors**

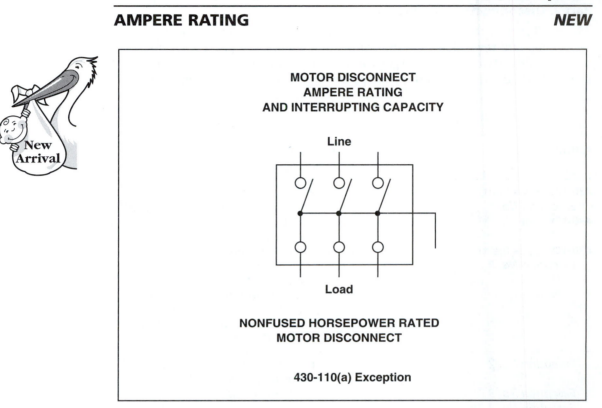

MOTOR DISCONNECT
AMPERE RATING
AND INTERRUPTING CAPACITY

Line

Load

NONFUSED HORSEPOWER RATED
MOTOR DISCONNECT

430-110(a) Exception

BASIC RULE

The motor disconnecting means for a 600-volt or less motor is required to have an ampere rating of not less than 115 percent of the full-load current rating of the motor.

NEW EXCEPTION

Exception: Listed non-fused motor circuit switches having a horsepower rating equal to or more than the horsepower rating of the motor are permitted to have an ampacity of less than 115 percent of the motor full-load current.

Reason

This exception recognizes a currently listed product. The switch is tested and listed for the horsepower without having the 115 percent ampere rating.

Comment

An isolating switch is not designed to be opened under load. The isolating switch is used to isolate a piece of electrical equipment after the circuit has interrupted by other means.

MOTOR DISCONNECT

COMBINATION LOADS

430-110(c)(1)

REVISED

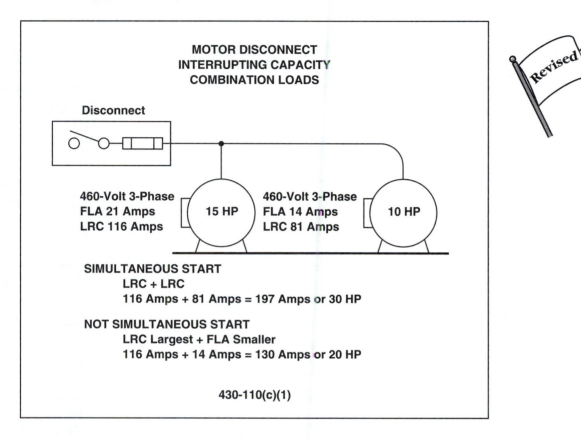

MOTOR DISCONNECT
INTERRUPTING CAPACITY
COMBINATION LOADS

Disconnect

460-Volt 3-Phase
FLA 21 Amps   **15 HP**
LRC 116 Amps

460-Volt 3-Phase
FLA 14 Amps   **10 HP**
LRC 81 Amps

SIMULTANEOUS START
    LRC + LRC
    116 Amps + 81 Amps = 197 Amps or 30 HP

NOT SIMULTANEOUS START
    LRC Largest + FLA Smaller
    116 Amps + 14 Amps = 130 Amps or 20 HP

430-110(c)(1)

REVISION

Where two or more motors or other loads cannot be started simultaneously, ~~appropriate combination of locked rotor and full-load current,~~ ***the largest sum of LRC of a motor or group of motors that can be started simultaneously and the full-load current of other concurrent loads*** shall be permitted to be used to determine the equivalent LRC for the simultaneous combination.

Reason

The term "appropriate combination" needed clarification.

CIRCUIT BREAKER

430-111(b)(2)

CONTROLLER AND DISCONNECT

REVISED

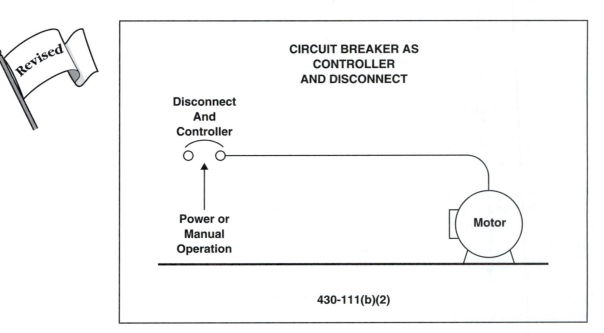

CIRCUIT BREAKER AS
CONTROLLER
AND DISCONNECT

430-111(b)(2)

REVISION

430-111 Switch or Circuit Breaker as Both Controller and Disconnecting Means This section is revised and reworded into a more readable outline with the addition of one new sentence: *When an inverse time circuit breaker is used as both the controller and the disconnecting means, it is permitted to be both power and manually operable.*

Cross Check

1996	1999

430-111 Switch or Circuit Breaker as Both Controller and Disconnecting Means
430-111 (part)
 	
(a) Air-Break Switch
(b) Inverse Time Circuit Breaker
(c) Oil Switch
	430-111 Switch or Circuit Breaker as Both Controller and Disconnecting Means
	(a) General
	(b) Type
	  (1) Air-Break Switch
	  (2) Inverse Time Circuit Breaker
	  (3) Oil Switch

GENERATORS   445-9

TERMINAL HOUSING   *NEW*

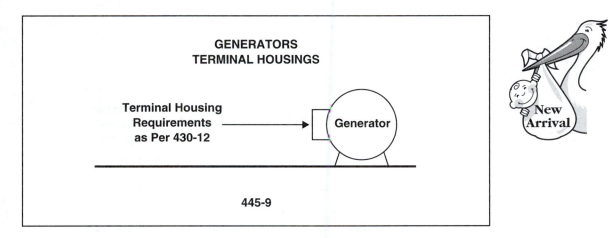

GENERATORS
TERMINAL HOUSINGS

Terminal Housing
Requirements ——————→ Generator
as Per 430-12

445-9

NEW RULE

445-9 Generator Terminal Housings *Generator terminal housings are now required to comply with Section 430-12.*

Reason

The revision recognizes that generator terminal housings are too small for the conductors that terminate in them. This new rule brings the sizing of generator terminal housings under the same requirements as motor terminal housings.

Comment

Section 430-12, Motor Terminal Housings, lists the minimum dimensions for terminal housings.

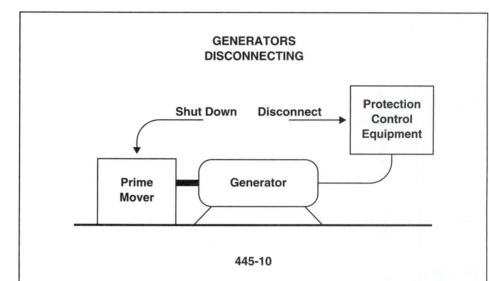

NEW RULE

445-10 Disconnecting Means Required for Generators *Generators shall be equipped with a disconnect, which disconnects the generator and all protective devices and control apparatus entirely from the generator except where:*

(a) The driving means for the generator can be readily shut down.

(b) The generator is not arranged to operate in parallel with another generator or other source of voltage.

Reason

Previously, there were no requirements for the supply side of an engine-driven generator. The new rule addresses the shut-down of the prime mover.

ARTICLE 450

TRANSFORMERS AND TRANSFORMER VAULTS (INCLUDING SECONDARY TIES)

The title of this article has a subheading added *(Including Secondary Ties).*

Section 450-3, Overcurrent Protection for Transformer, has undergone a major revision. The entire section is deleted and largely replaced with two tables. Some of the exceptions are deleted completely; others are deleted and made a part of the tables or the notes to the tables. The revision clarifies rules without changing them.

DELETIONS

~~450-3 Overcurrent Protection~~
~~(a) Transformers Over 600 Volts, Nominal~~
  ~~*All Exceptions*~~
  ~~Table 450-3(a)(1)~~
~~(b) Transformers 600 Volts, Nominal or Less~~
  ~~*All Exceptions*~~
  ~~Table 450-3(b)(2)~~

REPLACEMENT

450-3 Overcurrent Protection *Overcurrent protection of transformers is required to comply with (a), (b), or (c).*

(FPN No. 1) Coordinate with Article 240 (no change)

(FPN No. 2) Nonlinear load caution (no change)

(a) Transformers Over 600 Volts, Nominal *Overcurrent protection is required as specified in Table 450-3(a).*

Table 450-3(a) Maximum Rating or Setting of Overcurrent Protection for Transformers Over 600 Volts (as a percentage of transformer rated current).

New Arrival

TRANSFORMERS OVER 600 VOLTS

Table 450-3(a) – Minimum Rating or Setting of Overcurrant Protection for Transformers Over 600 volts (as a percentage of transformer rated current)

| Location Limitations | Transformer Rated Impedance | Primary Protection | | Secondary Protection (see Note 2) | | |
| | | Over 600 Volts | | 600 Volts or Below | | |
		Circuit Breaker (see Note 4)	Fuse Rating	Circuit Breaker (see Note 4)	Fuse Rating	Circuit Breaker or Fuse Rating
Any Location	Not more than 6%	600% (see Note 1)	300% (see Note 1)	300% (see Note 1)	250% (see Note 1)	125% (see Note 1)
	More than 6% and not more than 10%	400% (see Note 1)	300% (see Note 1)	250% (see Note 1)	225% (see Note 1)	125% (see Note 1)
Supervised Locations Only (see Note 3)	Any	300% (see Note 1)	250% (see Note 1)	Not Required		
	Not more than 6%	600%	300%	300%	250%	250%
	More than 6% and not more than 10%	400%	300%	250%	225%	250%

Table 450-3(a)

Cross Check

1996	1999
450-3(a)(1) *Exception No. 1* 450-3 (part) **450-3(a)(2) Supervised Installations** 450-3(a)(1) (part)	Notes to Table 430-3(a) **Note 1** Next size larger **Note 2** Six OCs permitted **Note 3** Supervised **Note 4** Electronic fuses **Note 5** Coordinated overloads (new)

(b) Transformers 600 Volts, Nominal, or Less *Overcurrent protection is required to be provided as specified in Table 450-3(b). Exception: Motor control transformers.*

Table 450-3(b) Maximum Rating or Setting of Overcurrent Protection for Transformers 600 Volts and Less (as a percentage of transformer rated current).

New Arrival

TRANSFORMERS
600 VOLTS OR LESS

Table 450-3(b) – Maximum Rating or Setting of Overcurrant Protection for Transformers 600 volts and less (as a percentage of transformer rated current)

Protection Method	Primary Protection			Secondary Protection (see Note 2)	
	Currents of 9A or more	Currents less than 9A	Currents less than 2A	Currents of 9A or more	Currents less than 9A
Primary only Protection	125% (see Note 1)	167%	300%	Not Required	
Primary and Secondary protection	250% (see Note 3)			125% (see Note 1)	167%

Table 450-3(b)

	Notes to Table 430-3(b)
450-3(b)(1) *Exception* (part) 450-3 (part) 450-3(b)(2)	**Note 1** Next size larger **Note 2** Six OCs permitted **Note 3** Thermal protection

(c) Voltage Transformer. Voltage transformers installed indoors are required to be protected by primary fuses.

450-4 Autotransformers 600 Volts, Nominal or Less

(a) Overcurrent Protection The exception is deleted, reworded, and made a part of the text with no change in the rule.

450-8 Guarding

(b) Case or Enclosure Dry-type transformers are required to be provided with a noncombustible moisture-resistant case or enclosure that provides ~~reasonable~~ protection against the accidental insertion of foreign objects.

Reason
The word "reasonable" could be interpreted variously. Its deletion now makes the rule enforceable.

Cross Check

1996	1999
450-13 Location (part of text) *Exception No. 1* **450-13 Location** (part of text) *Exception No. 2* (FPN No. 1) Moved to 450-21(b) (FPN No. 2) Deleted	**450-13 Accessibility** **(a) Open Installation** **(b) Hollow Space Installations** (Part of text)

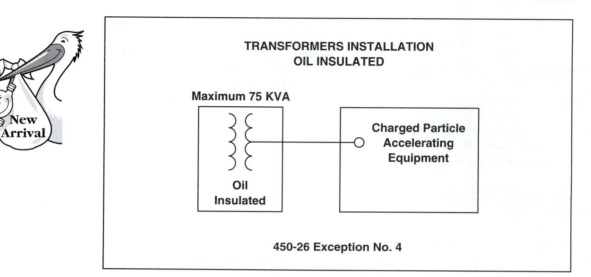

BASIC RULE

450-26 Oil-Insulated Transformers Installed Indoors When installed indoors, oil-insulated transformers are required to be installed in vaults.

NEW EXCEPTION

Exception No. 4: A transformer having a total rating not exceeding 75 kVA and a supply voltage of 600 volts or less that is an integral part of charged particle accelerating equipment shall be permitted to be installed without a vault in a building or room of noncombustible or fire-resistant construction, provided suitable arrangements are made to prevent a transformer oil fire from spreading to other combustible material.

Reason

It is not feasible to install an oil-filled transformer used for semiconductor production in a vault.

Comment

Semiconductor manufacturers use oil-filled transformers because of their low dielectric properties. The process temperatures are reasonably low, lessening the chance of an oil-caused fire.

    With the insertion of this exception, the existing Exceptions No. 4 and 5 now become Exceptions No. 5 and 6.

ARTICLE 455

PHASE CONVERTERS

This article has been rearranged into a more readable outline. It has been reworded and exceptions have been transferred into the text. Some new material is added.

PHASE CONVERTERS   455-6(b)

PHASE MARKINGS   *NEW*

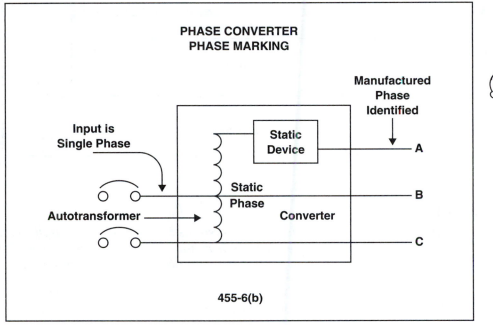

NEW RULE

The manufactured phase conductors are required to be identified at all accessible locations with a distinctive marking.

Reason
Identification of the manufactured phase is necessary at all points to ensure that no single-phase loads are connected to the manufactured phase.

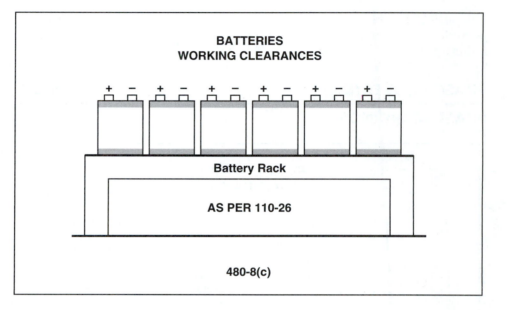

**BATTERIES
WORKING CLEARANCES**

Battery Rack

AS PER 110-26

480-8(c)

NEW RULE

The working space about the battery systems is required to comply with Section 110-26. The working clearance is required to be measured from the edge of the battery rack or containment area around the battery bank.

Reason

Sufficient working space is needed for proper maintenance.

Comment

Section 110-26 lists the various working spaces required around electrical equipment. The working space is required all around the battery bank; thus a battery bank on a rack cannot be placed against a wall.

EQUIPMENT OVER 600 VOLTS, NOMINAL   ARTICLE 490

NEW HOME   *MOVED*

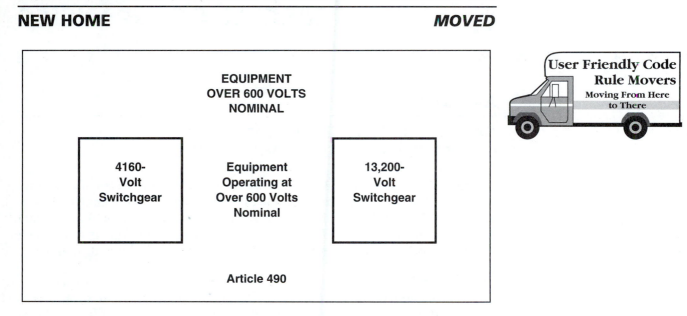

NEW ARTICLE

Article 710, Over 600 Volts, Nominal, is completely deleted in the 1999 *Code*. Most of the rules formerly in Article 710 are now in a new **Article 490, Equipment Over 600 Volts, Nominal.** Some sections of Article 710 are moved to other articles of the *Code* that have a part entitled "Over 600 Volts, Nominal." In making the change, many exceptions are deleted, reworded, and incorporated into the text.

Reason

The installation of the higher voltage is becoming more common; it is no longer a special condition. Already several articles have a part dedicated to "Over 600 Volts, Nominal."

CHAPTER 5
SPECIAL OCCUPANCIES

■ ■

ARTICLE 500

HAZARDOUS (CLASSIFIED) LOCATIONS
CLASS I, II, AND III, DIVISIONS 1 AND 2

Articles 500 through *504* are completely separated from *Articles 500-5, Class I, Zone 0, 1, and 2.* The title of Article 500 is changed to indicate what it covers *Class I, II, and III, Divisions 1 and 2.* All previous references to Article 505 throughout Articles 500, 501, 502, and 503 are deleted. An FPN is introduced in Section 500-1 Scope referencing Article 505 for Class I, Zone 0, 1, and 2.

New Section

500-3(b) Documentation. *All areas designed as hazardous (classified) locations shall be properly documented. This documentation is to be available to those authorized to design, install, inspect, maintain, or operate electrical equipment at the location.*

Reason

This new rule was a tentative interim amendment (TIA) and is now made a part of the *Code.*

CLASS I DIVISION 1 *NEW*

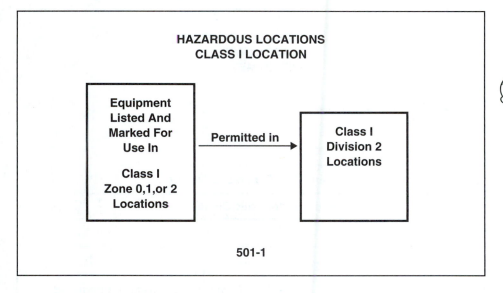

**HAZARDOUS LOCATIONS
CLASS I LOCATION**

Equipment Listed And Marked For Use In

Class I Zone 0,1,or 2 Locations

→ Permitted in →

Class I Division 2 Locations

501-1

NEW RULE

501-1 General. *Equipment listed and marked in accordance with Section 505-10 for use in Class I, Zone 0, 1, and 2 locations is permitted to be used in Class I, Division 2 Locations for the same gas and with a suitable temperature rating.*

Reason

Basically the *Code* treats Class I, Zone 1 and 2 as equivalent to Class I, Division 2. Therefore, the materials should be usable in either location.

Comment

Section 505-10 covers the listing, marking, and documentation for equipment to be used in Class I, Zone 0, 1, or 2.

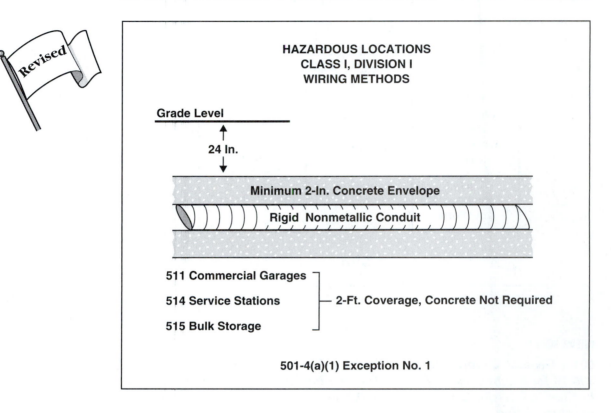

**HAZARDOUS LOCATIONS
CLASS I, DIVISION I
WIRING METHODS**

Grade Level

24 In.

Minimum 2-In. Concrete Envelope

Rigid Nonmetallic Conduit

511 Commercial Garages

514 Service Stations   — 2-Ft. Coverage, Concrete Not Required

515 Bulk Storage

501-4(a)(1) Exception No. 1

BASIC RULE

The wiring methods in Class I, Division 1 are limited to rigid metal conduit, intermediate metal conduit, and MI cable.

REVISED EXCEPTION

The exception permitted rigid nonmetallic conduit encased in 2 in. concrete, buried below the surface under not less than 2 ft of earth *and provided with not less than 24 in. of cover measured from the top of the encasement to grade of an underground run. Concrete is permitted to be omitted where not required in Section 511-4 Exception, 514-8 Exception No. 2, and 515(a).* Where nonmetallic conduit is used, the last 24 in. must be rigid metal conduit or intermediate conduit. An equipment grounding conductor is required.

Reason

The change in the wording clarifies the depth of the burial. The reference to the installations without concrete encasement coordinates with other sections of the *Code*.

HAZARDOUS LOCATIONS   501-4(a)(1) Exception No. 3

CLASS I, DIVISION 1 WIRING METHODS   *NEW*

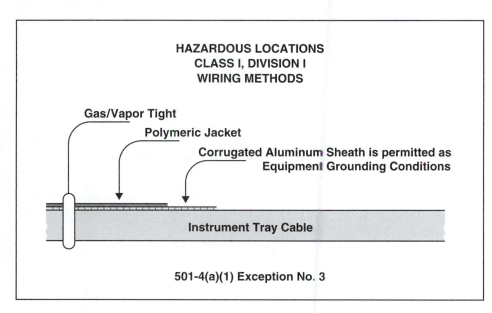

**HAZARDOUS LOCATIONS
CLASS I, DIVISION I
WIRING METHODS**

Gas/Vapor Tight

Polymeric Jacket

Corrugated Aluminum Sheath is permitted as
Equipment Grounding Conditions

Instrument Tray Cable

501-4(a)(1) Exception No. 3

New Arrival

BASIC RULE

The wiring methods in Class I, Division 1 are limited to rigid metal conduit, intermediate metal conduit, and MI Cable.

NEW EXCEPTION

Instrument tray cable is permitted as a wiring method in a Class I, Division 1, location provided all the following conditions are met.
* *In industrial establishments with restricted public access,*
* *Qualified maintenance and supervision ensured.*
* *Cable is gas/vapor tight.*
* *Continuous corrugated aluminum sheath*
* *Suitable polymeric jacket*
* *Aluminum sheath is permitted as the equipment grounding conductor.*
* *Terminate in listed fittings.*

Reason

Like MC Cable, ITC cable is constructed with 300-volt conductor insulation. Type MC Cable is permitted in Exception No. 2, Article 727, covers Type ITC Cable and limits its use to 150 volts.

CLASS I, DIVISION 1 COMPOUND THICKNESS   *REVISED*

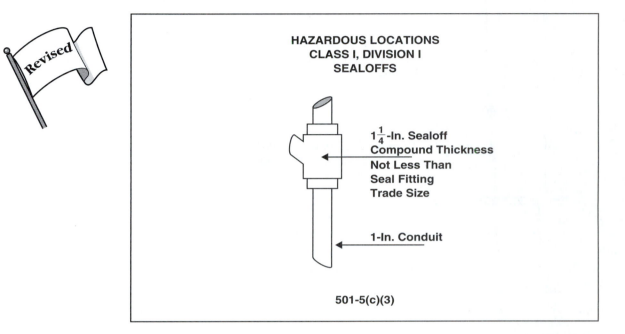

HAZARDOUS LOCATIONS
CLASS I, DIVISION I
SEALOFFS

$1\frac{1}{4}$-In. Sealoff
Compound Thickness
Not Less Than
Seal Fitting
Trade Size

1-In. Conduit

501-5(c)(3)

REVISION

In a completed seal, the minimum thickness of the sealing compound is required to be not less than the trade size of the ~~conduit~~ *sealing fitting,* and in no case less than $5/8$ in.

Reason

By deleting the word "conduit" and replacing it with "sealing fitting," the rule now covers both conduit and cable installations. This also helps to compensate for the 25 percent conductor fill limitation for a sealoff fitting.

REVISION

501-5 Sealing and Draining.

(c) Class I, Division 2 and 2.(6) Conductor Fill The cross-sectional area of conductors permitted in a seal shall not exceed 25 percent of the cross-sectional area of a *rigid metal* conduit of the same trade size unless it is specifically approved for a higher percentage of fill.

Reason

The fill limitation for a sealoff is based on standard sizes of rigid metal conduit.

501-5(d) Cable Seals, Class I, Division 1.

Basic rule: (1) Each individual conductor is required to be separated and sealed around.

   New Exception: *Shielded cables and twisted pair cables are permitted to be installed without separating the conductors.*

Reason

Untwisting the twisted pairs defeats the purpose of twisting the cable to lessen noise.

Note

This same exception is made in Section 501-5(e)(1) for Class I, Division 2, locations.

HAZARDOUS LOCATIONS   501-5(e)(1) Exception No. 1

CLASS I, DIVISION 2   *NEW*

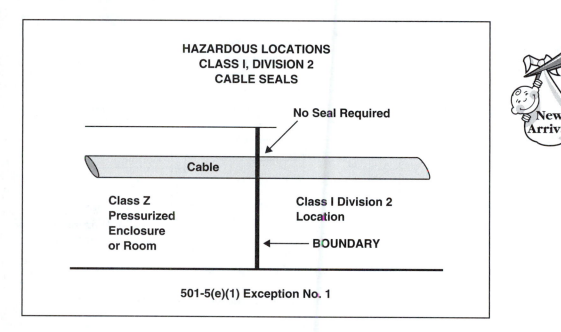

HAZARDOUS LOCATIONS
CLASS I, DIVISION 2
CABLE SEALS

No Seal Required

Cable

Class Z
Pressurized
Enclosure
or Room

Class I Division 2
Location

BOUNDARY

501-5(e)(1) Exception No. 1

New Arrival

BASIC RULE

Cables are required to be sealed at the point of entry.

NEW EXCEPTION

Cables passing from an enclosure or room that is unclassified as a result of Type Z pressurization into a Class I, Division 2 location is not required to be sealed at the boundary.

Reason

This new exception explains the rule more clearly.

DELETION

Section 501-6(b)(4) ~~Fuses or Circuit Breakers for Overcurrent Protection~~
This section limited the number of circuit breakers or sets of fuses for circuits to 10.

Reason

There was no reason to limit the number of circuits.

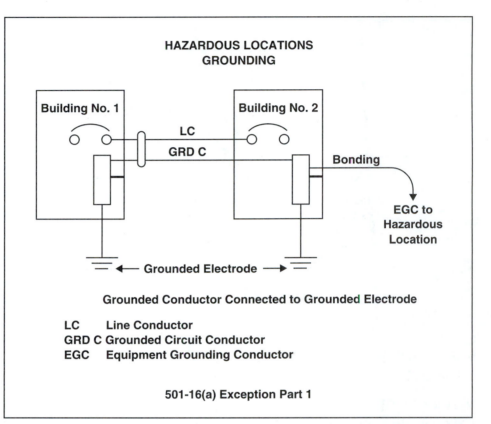

**HAZARDOUS LOCATIONS
GROUNDING**

Building No. 1

Building No. 2

LC

GRD C

Bonding

EGC to
Hazardous
Location

◄— Grounded Electrode —►

Grounded Conductor Connected to Grounded Electrode

LC   Line Conductor
GRD C Grounded Circuit Conductor
EGC   Equipment Grounding Conductor

501-16(a) Exception Part 1

BASIC RULE

In hazardous locations fittings with metal raceways should not be relied on for continuity of the equipment grounding conductor. The continuity must be maintained by using bonding jumpers back to the point of grounding for the service.

REVISED EXCEPTION

The specific bonding means is only required to the point *where the grounded circuit conductor and the grounding electrode are connected together on the line side of the* ~~of grounding of a~~ building disconnecting means as specified in Section 250-24 32(a), (b), and (c), provided the branch-circuit overcurrent protection is located on the load side of the disconnecting means.

Reason

The revision protects and insures the continuity of the equipment grounding path for hazardous locations.

Comment

Section 250-32 establishes rules for grounding at a separate building or structure. Illustration part 1 is based on Section 250-32(a) with no equipment grounding conductor installed to the second building. The bonding is only required back to the separate building supply where the grounded circuit conductor is connected to a grounding electrode.

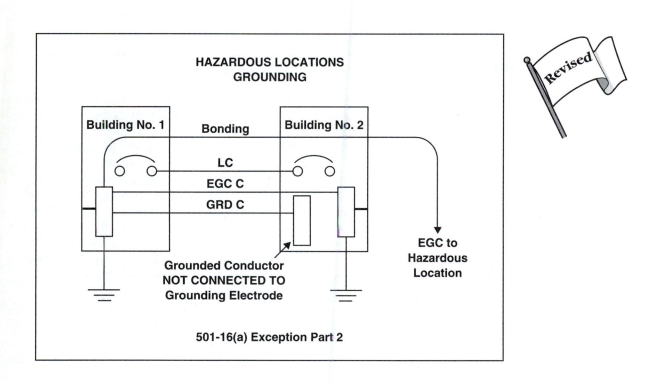

HAZARDOUS LOCATIONS GROUNDING

Building No. 1   Bonding   Building No. 2

LC

EGC C

GRD C

Grounded Conductor NOT CONNECTED TO Grounding Electrode

EGC to Hazardous Location

501-16(a) Exception Part 2

In part 2 of the illustration, the installation is in accordance with Section 250-32(b)(1). An equipment grounding conductor is installed back to the first building, and there is no connection between the grounded conductor and the grounding electrode at the second building. Therefore, the bonding is required back to the point where the grounded conductor is connected to the grounding electrode in the originating building.

ARTICLE 505

CLASS I, ZONE 0, 1, AND 2 LOCATIONS

This article was first introduced into the *Code* in the 1996 edition. Now it is completely rearranged and updated, with the insertion of new material and an expansion of existing materials.

Many revisions are linked to the "Protective Techniques" introduced in Section 505-4. Some *Code* section numbers change, while others remain the same. There is extensive use of Fine Print Notes for referencing other codes, explanation, examples, and special information. A number of new symbols and terms are introduced and either defined or explained as they are used.

Section 505-6(a) requires the installation of Article 505, Class I, Zone 0, 1, and 2, to be under the supervision of a registered professional engineer.

Section 505-21 is a new section recognizing and giving the rules for the installation of ***Increased Safety "e" Motors and Generators***. These are also referred to as Design B Energy Efficient motors.

ARTICLE 513

AIRCRAFT HANGERS

513-1 Scope

A new section **513-1 Scope** is introduced at the beginning of Article 513, moving all other sections back by one digit. The new scope indicates what this article is intended to cover and is followed with two new Fine Print Notes.

FPN No. 1 references "Standards on Aircraft Hangers" for applicable definitions.

FPN No. 2 references documents for fuel classification.

The former Section 513-1, Definition, is deleted and replaced with a new Section **513-2, Definitions,** with separate definitions for **Portable Equipment** and **Mobile Equipment.**

Reason

A scope was needed for the article. The definition used in the 1996 *Code* did not accord with the definitions given in Standards on Aircraft Hangers, NFPA 406-1995. The definitions and the new FPN will more closely align the two *Codes*.

513-4 Wiring and Equipment in Class I Locations

Delete the last part of the sentence. Where such wiring is located in vaults, pits, or ducts, adequate drainage shall be provided. ~~and the wiring shall not be placed within the same compartment with any service other than piped compressed air.~~

Reason

There seemed to be no obvious reason for this restriction.

ARTICLE 514

GASOLINE DISPENSING AND SERVICE STATIONS

A new section is inserted for **514-6** and **514-6 Sealing** of the 1996 *Code* is now **514-7**.

    514-6. *Provisions for Maintenance and Service of Dispensing Equipment. Each dispensing device is now required to be provided with a means to remove all external voltage sources, including feedback, during periods of maintenance and service of dispensing equipment.*

Reason
Installation methods that do not isolate remote pump control wiring promote feedback in the wiring and are hazardous to personnel and equipment.

ARTICLE 517

HEALTH CARE FACILITIES

The many FPNs used to reference the "Life Safety *Code*," "Standards for Health Care Facilities," and other documents are updated to the most current edition of these documents. For example, the "Life Safety Code," NFPA 101, is updated from the 1994 to 1997 edition, and the "Standards for Health Care Facilities," NFPA 99, from the 1993 to the 1996 edition. Rules taken from these documents are identified by a superscript "X" and specifically identified by cross checking the section number in Appendix A. The many changes indicated in the FPNs occur because of references to new editions.

PATIENT CARE AREA EQUIPMENT GROUNDING   *REVISED*

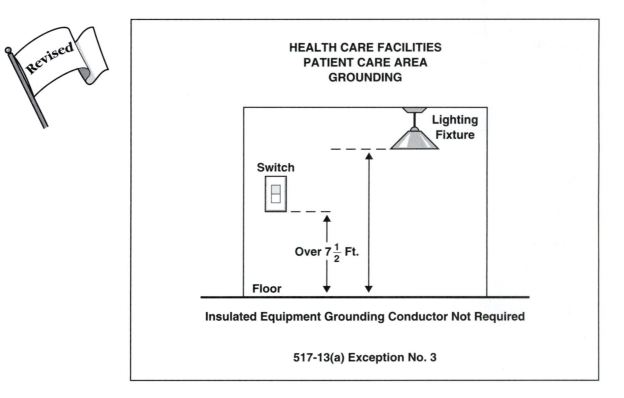

HEALTH CARE FACILITIES
PATIENT CARE AREA
GROUNDING

Lighting Fixture

Switch

Over $7\frac{1}{2}$ Ft.

Floor

Insulated Equipment Grounding Conductor Not Required

517-13(a) Exception No. 3

BASIC RULE

All electrical equipment, including metal faceplates of switches and receptacles are required to be grounded using an insulate equipment grounding conductor.

REVISED EXCEPTION

The words *"switch located outside the vicinity"* are added to the exception, making it apply to both the lighting fixture and switches mounted over $7^1/_2$ ft above the floor.

Reason

This revision clarifies the requirements for grounding switches installed in the patient care area.

HEALTH CARE FACILITIES   517-18(a) Exception No. 3

PATIENT BED LOCATION BRANCH CIRCUITS   *NEW*

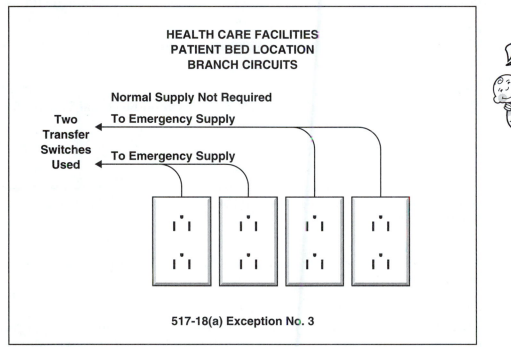

**HEALTH CARE FACILITIES
PATIENT BED LOCATION
BRANCH CIRCUITS**

Normal Supply Not Required

Two Transfer Switches Used

To Emergency Supply

To Emergency Supply

517-18(a) Exception No. 3

BASIC RULE

The patient bed location is required to be supplied by at least two branch circuits, one or more from the emergency system, and one or more from the normal system.

NEW EXCEPTION

Where a patient bed location is served from two separate transfer switches on the emergency system, a circuit from the normal system is not required.

Reason

This exception extends the same consideration to the General Care Area as to the Critical Care Area in Section 517-19(a) Exception No. 2.

517-19 Critical Care Area

(b) Patient Bed Location Receptacles This subsection is revised and outlined so that it agrees with the preceding section about the use of two transfer switches. The following is a brief outline.

(b) Patient Bed Location Receptacles
    (1) 6 required.
        (a) 1 connected to normal system
        (b) or emergency supplied by different transfer switches
    (2) Single, duplex, or both.
        Listed hospital grade
        Insulated copper equipment grounding conductor

517-30 Essential Electrical Systems for Hospitals

(b) General Add the following:

(6) Hospital power sources and alternate power sources shall be permitted to serve the essential electrical systems of contiguous or same site facilities.

Reason

This statement is taken from NFPA 90. It clarifies what is permitted to be served.

Comment

This is also coordinated with Section **517-40, Essential Electrical Systems for Nursing Homes and Limited Care Facilities.**

517-33 Critical Branch

(a) Task Illumination and Selected Receptacles

(9) Single-phase fractional horsepower ~~exhaust fan motors that are interlocked with 3-phase motors on the equipment~~ shall be permitted to be connected to the critical branch.

Reason

This deletion brings this section into agreement with NFPA 99, "Standards for Health Care Facilities."

INSERT NEW DEFINITION

517-33(c)^X Receptacle Identification. The receptacle or the faceplates for receptacles supplied by the critical branch shall have a distinctive color or marking so as to be readily recognizable.

Reason

This definition is needed. The superscript X indicates that it is taken from another document, which is identified in Appendix A. In this case, the document is NFPA 99, "Standards for Health Care Facilities."

ISOLATED TRANSFORMER—RECEPTACLE   *NEW*

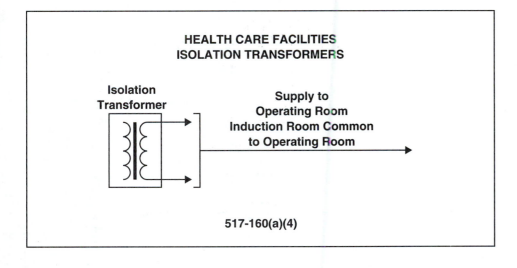

**HEALTH CARE FACILITIES
ISOLATION TRANSFORMERS**

Isolation Transformer

Supply to
Operating Room
Induction Room Common
to Operating Room

517-160(a)(4)

NEW RULES

This subsection is revised, rearranged, and new rules added.

(4)ˣ **Isolation Transformer**. An isolating transformer is not permitted to service more than one operating room *except as covered in a. and b. For the purpose of this section, anesthetic induction rooms are considered part of the operating room or rooms or served by the induction room.*

a. Induction Rooms. Where an induction room serves more than one operating room, the isolated circuits of the induction room are permitted to be supplied from the isolation transformer of any of the operating rooms served by the induction room.

b. Higher Voltages. Isolating transformers are permitted to serve single receptacles in several patient areas where:

   1. Reserved receptacles serve equipment requiring 150 volts or higher: X-ray units.

   2. Receptacle mating plugs not interchangeable with other isolated power system.

Reason

Clarification of this section was needed. The revision includes regulations based on NFPA 99, "Standards for Health Care Facilities."

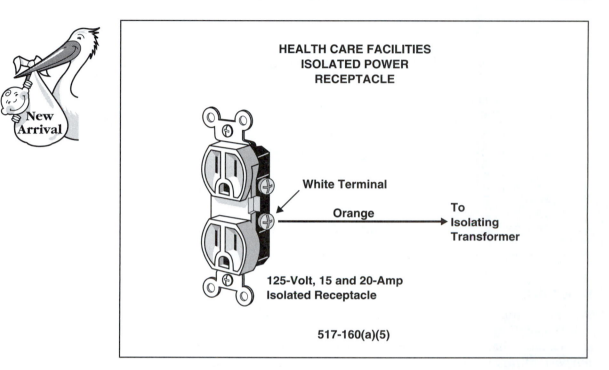

HEALTH CARE FACILITIES
ISOLATED POWER
RECEPTACLE

White Terminal

Orange → To Isolating Transformer

125-Volt, 15 and 20-Amp
Isolated Receptacle

517-160(a)(5)

BASIC REQUIREMENT

Where an isolating transformer is used, isolation conductor No. 1 is Orange and isolation conductor No. 2 is Brown.

REVISION

Where there is a 3-phase isolation transformer, the color code for the third wire is yellow.

Where isolated circuit conductors supply 125-volt single-phase, 15- and 20-ampere receptacles, the orange conductor(s) are required to be connected to the terminal(s) on the receptacles that are identified for the connection to the grounded-circuit conductor.

Reason

This coordinates with the Canadian Electrical Code, making for a standard application for this installation.

ARTICLE 518

PLACES OF ASSEMBLY

Article 518 is rearranged with more boldface headings. Some exceptions are deleted, reworded, and made a part of the text without a change in the intent of the rule.

Cross Check

1996	1999
518-1 Scope	**518-1 Scope**
518-2 General Classifications	**518-2 General Classifications** (title only)
Part of text	**(a) Examples**
Part of text	**(b) Multiple Occupancies**
	(c) Theatrical Areas
518-3 Other Articles	**518-3 Other Articles**
(a) Hazardous (Classified) Areas	**(a) Hazardous (Classified) Areas**
(b) Temporary Wiring	**(b) Temporary Wiring**
Exception No. 1	Part of text
Exception No. 2	Part of text
NEW	*Exception*
(c) Emergency Systems	**(c) Emergency Systems**
	518-4 Wiring Methods (title only)
518-4 Wiring Methods	**(a) General** (Part of text)
	Add Liquidtight Flexible Metal Conduit
Exception No. 2	*Exception* (revise wording)
Exception No. 1	**(b) Non Rated Construction**
(FPN)	(FPN)
Exception No. 3	**(c) Spaces with Finish Rating**
a.	**(1)**
b.	**(2)**
NEW	(FPN)
518-5 Supply	**518-5 Supply**

PLACES OF ASSEMBLY

WIRING METHODS

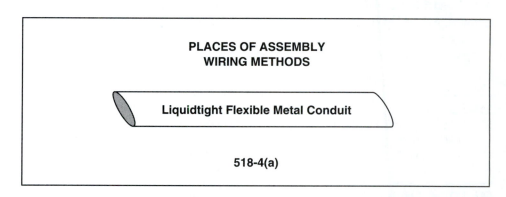

**PLACES OF ASSEMBLY
WIRING METHODS**

Liquidtight Flexible Metal Conduit

518-4(a)

REVISION

Flexible metal raceways are now permitted to be used in places of assembly.

Reason

This revision makes it clear that liquidtight flexible metal conduit and other flexible metal raceways are permitted to be used in places of assembly.

ARTICLE 520

THEATER, AUDIENCE AREAS, OR MOTION PICTURE AND TELEVISION STUDIOS, AND SIMILAR LOCATIONS

NEW DEFINITION

Breakout Assembly: An adapter used to connect a multipole connector containing two or more branch circuits to multiple individual branch circuit connectors.

Reason

The term "breakout assembly" used in the article was not defined.

520-25 Dimmers

(d) Solid State Dimmers. Delete the word ~~approved~~ and insert the word listed.

Reason

Deleting the word "approved" relieves the AHJ of having to make the decision about appropriate products. Using listed products adds to the safety of the installation.

TITLE CHANGE

1996 Title	Part C. Stage Equipment—Fixed
1999	**Part C. *Fixed Stage Equipment Other Than Switchboards***

520-53 Construction and Feeders

(h) Supply Conductors (2) Single Conductor Cables Delete and replace one sentence in center of paragraph. ~~The single conductor cables for a supply shall be of the same type, size, length and be grouped together, but not bundled.~~

Replace with: *Where single conductors are paralleled for increased ampacity, the paralleled conductors shall be of the same length and size. Single-conductor supply cables shall be grouped together but not bundled.*

Reason

The revision clarifies the rule and eliminates a conflict with another section.

NEW SECTION UNDER 520-53

(k) Single-Pole Separable Connectors. Add a new sentence for clarification: *Sections 400-10 and 410-56 shall not apply to listed single-pole separable connectors and single-conductor cable assemblies utilizing listed single-pole separable connectors.*

Comment

Section 400-10 requires flexible cords and cables to be connected to devices and fittings in such a way that tension is not transmitted to joints or terminals. Section 410-56 concerns the ratings and types of receptacles.

520-69 Adapters.

A new subsection is added.
(c) Conductor Type. Conductors for adapters and two-fers shall be listed extra-hard usage or listed hard usage (junior hard service) cord. Hard usage (junior hard service) cord shall be restricted in overall length to 3.3 ft. (1 m).

Reason

This revision coordinates the *Code* with a type of installation safely used in the field.

LIGHTS AND RECEPTACLE SWITCHING
DRESSING ROOMS　　　　*DELETED & REVISED*

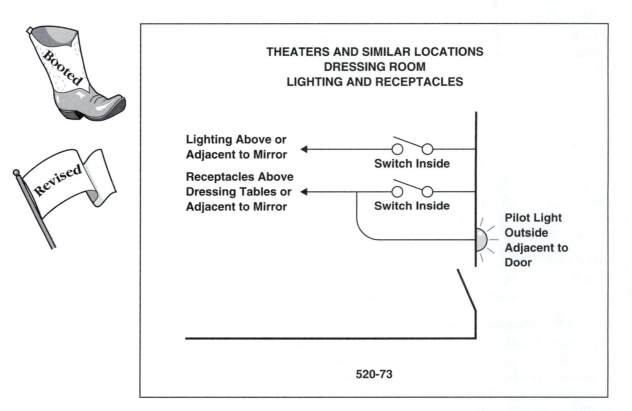

THEATERS AND SIMILAR LOCATIONS
DRESSING ROOM
LIGHTING AND RECEPTACLES

Lighting Above or
Adjacent to Mirror

Switch Inside

Receptacles Above
Dressing Tables or
Adjacent to Mirror

Switch Inside

Pilot Light
Outside
Adjacent to
Door

520-73

DELETION

~~All lights and receptacles in dressing rooms shall be controlled by wall switches installed in the dressing rooms. Each switch controlling receptacles is required to have a pilot light.~~

REVISION

All lights *adjacent to the mirrors and above dressing table counters* are required to be controlled by a switch in the dressing room. All receptacles *adjacent to mirrors and above the dressing table counter* are required to be controlled by a switch in the dressing room, and a pilot light is required *to be located outside the dressing room adjacent to the door* to indicate when the receptacles are energized. *Other outlets in the dressing room are not required to be switched.*

Reason

A modern dressing room is almost like an office, and it needs some receptacles with an uninterrupted power supply.

CARNIVALS, CIRCUSES, SIMILAR LOCATIONS 525-3(c)

SOUND EQUIPMENT *NEW*

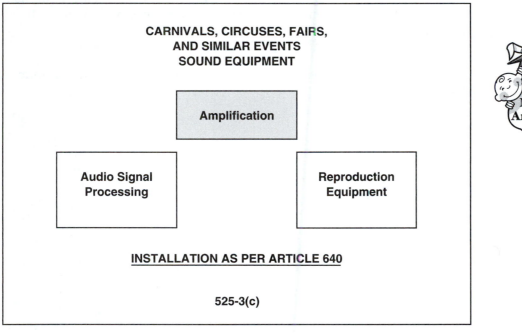

NEW RULE

525-3(c) Audio Signal Processing, Amplification and Reproduction Equipment. Article 640 shall apply to the wiring and installation of audio signal processing, amplification and reproduction equipment.

Reason

Traveling shows, such as folk festivals and rock concerts, utilize this sound equipment. This use was not previously covered in this article.

525-10 Power Sources

This section is rearranged so that separately derived systems, such as generators, come ahead of services.

Reason

Most events covered by this article do not have a service but have a separately derived system from a generator. A generator supplied system is not considered as a service-supplied system.

525-12(a) Vertical Clearances

A new sentence is added to this section: *The clearances required by this section apply to wiring installed outside of a tent and not to the wiring installed inside the tent or concessions.*

Reason

The tent roof is not as high as the outdoor clearances required.

ADD A NEW PART TO ARTICLE 525

E. Attractions Utilizing Pools and Fountains

525-40 Wiring and Equipment Associated with Pools and Fountains and Similar Installations with Contained Volume of Water. This equipment shall be installed to comply with the applicable requirements of Article 680.

Reason

There are attractions at these events that do have pools or fountains. Bumper boats is one example.

CARNIVALS, CIRCUSES, AND SIMILAR LOCATIONS   525-18

GFCI REQUIREMENTS   *DELETED & REVISED*

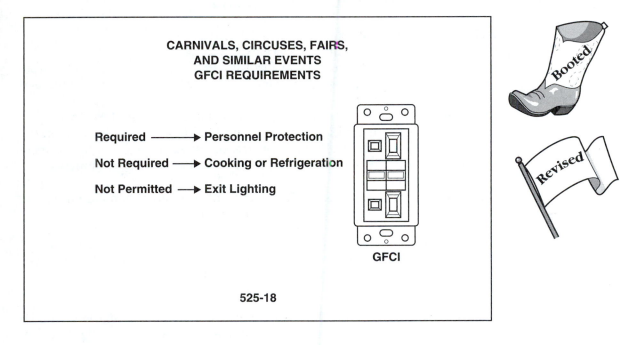

DELETION

~~The ground-fault-circuit-interrupter requirements of Section 305-6 shall not apply to this article.~~

REVISION

525-18 Ground-Fault Circuit-Interrupter Protection for Personnel.

(a) All 125 volts, 15- and 20-ampere receptacle outlets that are in use by personnel are required to have listed ground-fault protection for personnel. The ground-fault protection is permitted to be an integral part of the attachment plug, or located in the supply cord within 12 in. (305 mm) of the attachment plug. Egress lighting is not permitted on the load side of a GFCI protected receptacle.

(b) Receptacles supplying cooking or refrigeration equipment are not required to have GFCI protection.

(c) Other receptacles are permitted to have GFCI protection or a written procedure for an assured equipment grounding program as described in Section 305-6(b)(2) and continually enforced by a designated person.

Reason

The previous *Code* overlooked the protection of personnel. The revision makes for a much safer installation. The concern for cord- and plug-connected cooking equipment on GFCI receptacles has been solved. Section 305-6(b) covers the rule for using an assured equipment grounding system.

ARTICLE 530

MOTION PICTURE AND TELEVISION STUDIOS AND SIMILAR LOCATIONS

530-12 Portable Wiring.
Revise a long paragraph into subsections.

(a) Stage Set Wiring. Hard usage flexible cords and cables are to be used. Where subject to physical damage, extra hard usage flexible cords and cables are required.

(b) Stage Effects and Electrical Equipment Used as Stage Properties. The use of single- or multiconductor flexible cords and cables is permitted when protected from physical damage and secured by cable ties or insulated staples. The cords may be tap or circuits not over 20-amperes with listed devices.

(c) **Other Electrical Equipment.** Cords and cables of other than hard usage are permitted when supplied as a part of a listed assembly.

Reason
The rearrangement separates cords and cables into their respective uses. The word "approved" is deleted, thereby assisting the AHJ.

Comment
Stage lighting includes floodlights, spotlights, and other such lighting. Special effect lighting may be outline lighting or power for special effects lighting controls for animations. Subsection (c) permits table lamps and other such appliances with their factory listed cord or cable to be used on stage.

MOTION PICTURE STUDIO

PORTABLE FEEDER CABLE

530-18(c)

NEW

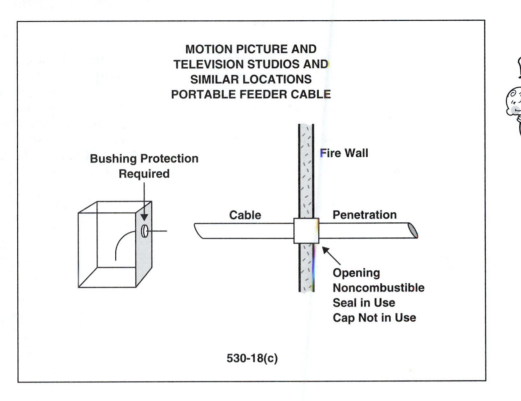

**MOTION PICTURE AND
TELEVISION STUDIOS AND
SIMILAR LOCATIONS
PORTABLE FEEDER CABLE**

Fire Wall

Bushing Protection
Required

Cable   Penetration

Opening
Noncombustible
Seal in Use
Cap Not in Use

530-18(c)

NEW RULES

Portable cables are required to be protected by bushings where they pass through enclosures and arranged to protect against tension on cable.

Portable feeders are permitted to temporarily penetrate fire rated walls, floor or ceilings provided:
(1) the opening be of noncombustible material
(2) When in use, the penetration is sealed with a temporary seal of a listed fire stop materials.
(3) When not in use, the opening is required to be capped with a material of equivalent fire rating.

Reason

There is a need for this new rule because old buildings are being converted to sound stages and there are materials on the market to fulfill the rule. Incidents have been observed where cables were installed through knockouts in panels and through fire walls with no protection.

ARTICLE 540

MOTION PICTURE PROJECTORS

540-11 Location of Associated Electrical Equipment.

Revise and add new material to subsection (a):

(a) Motor Generator Sets, Transformers, Rectifiers, Rheostats, and Similar Equipment. Motor generator sets, transformers, rectifiers, rheostats, and similar equipment for the supply or control of current to projection or spotlight equipment shall, if practical, *where nitrate film is used*, be located in a separate room. Where placed in the projection room, they shall be so located or guarded that arcs or sparks cannot come in contact with film, and *the end or ends* of motor generator sets shall *comply with one of the conditions in (1) through (6) below.*

(1) Types.

(2) Separate rooms or housings.

(3) Solid metal cover.

(4) Tight metal housings.

(5) Upper and lower half enclosures.

(6) Wire screens or perforated metal.

Reason

The revision clarifies the rule and improves safety.

ARTICLE 547

AGRICULTURAL BUILDINGS

547-4 Wiring Methods

This section is rearranged and reworded.

Delete the redundant statement *In agricultural buildings* as described in Section 547-1(a) and (b) in several places as it is covered in the scope of the article. The sections are moved because they apply to the wiring methods. The material is reworded into complete sentences without changing the rules.

Cross Check

1996	1999

547-4 Wiring Methods
    FPN
547-4 Wiring Methods
  (a) Boxes, Fittings, and Wiring Devices
  (b) Flexible Connections
547-10 Physical Protection
547-8(c) Separate Equipment Grounding Conductor
	547-4 Wiring Methods (Title only)
	  (a) Wiring Systems (revised)
	      FPN
	  (b) Mounting (Part of text)
	  (c) Boxes and Fittings
	  (d) Flexible Connections
	  (e) Physical Protection
	  (f) Separate Equipment Grounding Conductors.

	547-8 Service Equipment, Separately Derived Systems, Feeders, Disconnecting Means, and Grounding New
	  (a) Disconnecting Means and Overcurrent Protection at Building(s) New
	  (b) Disconnecting Means and Overcurrent Protection at Distribution Point New
	  (c) Disconnect Means Without Overcurrent Protection at the Distribution Point.
	      (1), (2), (3), (4), and (5)
	  Distribution Point New definition

547-8 Grounding, Bonding, and Equipotential Plane
  Equipotential Plane (definition)
  (b) Concrete Embedded Elements
 	
  (b) FPN
  (c) FPN
	547-9 Bonding and Equipotential Plane
	
	  (a) Definition Revised
	  (b) General
	      Exception No. 1 New
	      Exception No. 2 New
	      FPN No. 1 New
	      FPN No. 2
	      FPN No. 3
	  (c) Receptacles New

Sections 547-5, 6, and 7 remain with no change in the intent of the rules.	

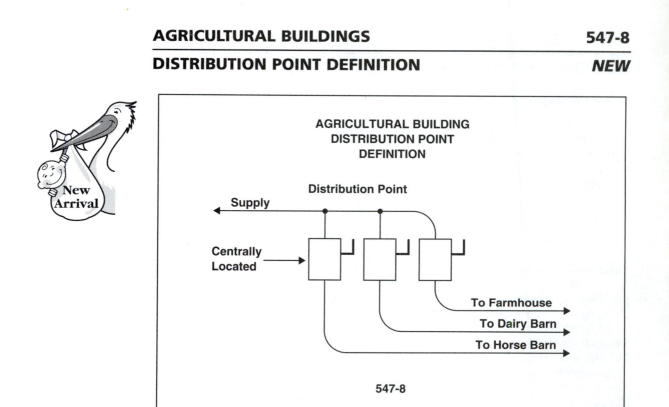

New Arrival

AGRICULTURAL BUILDING
DISTRIBUTION POINT
DEFINITION

Distribution Point

Supply

Centrally
Located

To Farmhouse
To Dairy Barn
To Horse Barn

547-8

NEW DEFINITION

A distribution point is a centrally located electrical supply structure from which services or feeders to agricultural buildings and other buildings, including the associated farm house, are normally supplied.

Reason
The term "distribution point," used in a new section of Article 547, Agricultural Buildings, requires definition.

Comment
A distribution point is very much like what is often referred to as a "load center" in the field. When NO overcurrent protection is located at the distribution point, only a disconnecting means, the conductors going to the various locations are service conductors. When it has both overcurrent protection and disconnecting means located at the distribution point, the conductors going to the various locations are feeders.

AGRICULTURAL BUILDINGS   547-8(a)

OVERCURRENT PROTECTION AT BUILDING   *NEW*

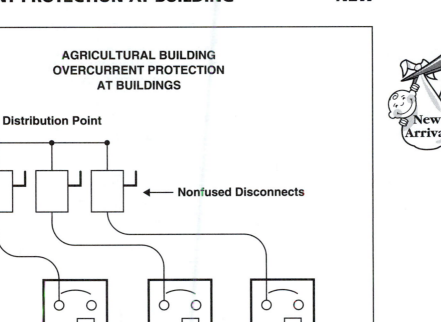

AGRICULTURAL BUILDING
OVERCURRENT PROTECTION
AT BUILDINGS

547-8(a)

NEW RULE

(a) Disconnect and Overcurrent Protection at Building(s). Where a disconnecting means and overcurrent protection are located at the load end of the service conductors, grounding shall meet the requirements of Section 250-24. Where two or more buildings are supplied from the distribution point, a disconnecting means is required at the distribution point for each building.

Reason
The recognition of the distribution point necessitated this change.

Comment
The conductors arriving at the separate buildings are service conductors, and the termination is treated as service conductors. The equipment grounding of the nonfused disconnects located at the distribution point is accomplished by bonding the disconnects to the grounded conductor as the disconnects are equipment on the line side of the service overcurrent protection.

OVERCURRENT PROTECTION AT DISTRIBUTION POINT *NEW*

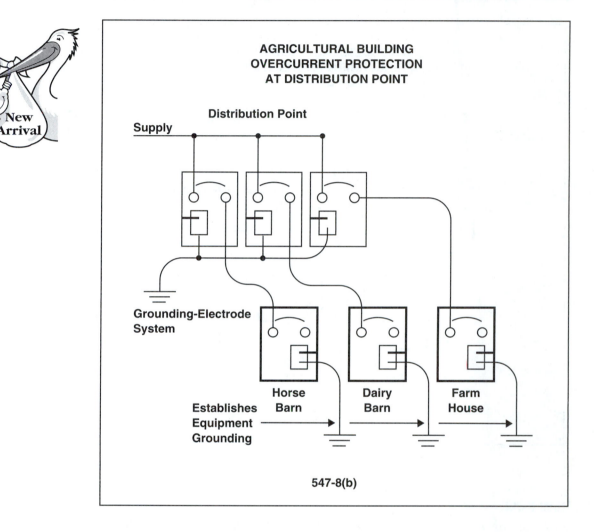

NEW RULE

(b) Disconnect and Overcurrent Protection at Distribution Point. Where a disconnection means and overcurrent protection are located at the distribution point, service conductors or feeders to buildings shall meet the requirements of Article 225 Part B and 250-32.

Reason
The recognition of the distribution point necessitated this change.

Comment
Part B of Article 225 covers the supply to a second building. By making this reference, it treats the distribution point as a building, and thus the agricultural buildings are being supplied by feeders via the distribution point.

Section 250-32 covers the installation of the supply at a separate building or structure.

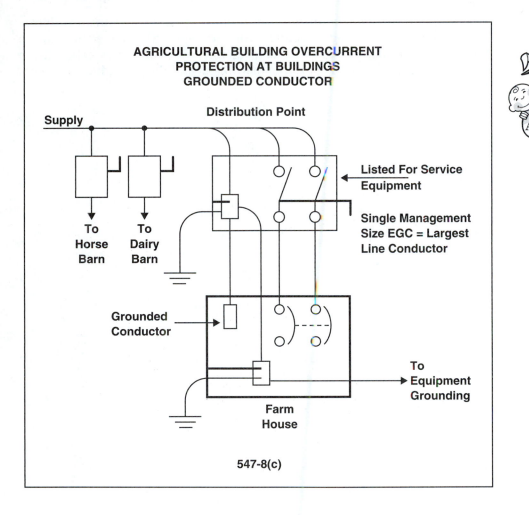

NEW RULE

(c) Disconnect Only at the Distribution Point. There is no overcurrent protection at the distribution point. There is overcurrent protection at the building. The grounded conductor is not required to be connected to the grounding conductor at the building provided all the following conditions are met:

(1) All under one management

(2) Listed service switch at distribution point

(3) Equipment grounding conductor (EGC) is run with circuit conductors and is at least as large as the largest service conductor.

(4) Grounded conductor is bonded to the grounding conductor at the distribution point.

(5) A grounding electrode is connected to the equipment grounding conductor at the building.

Reason

The recognition of the distribution point necessitated this change.

Comment

The term "grounding conductor" as used here indicates the conductor between the nonfused disconnecting switch at the distribution point and the grounding electrode. It is also used to identify the conductor between the equipment ground bar at the building disconnect and the grounding electrode.

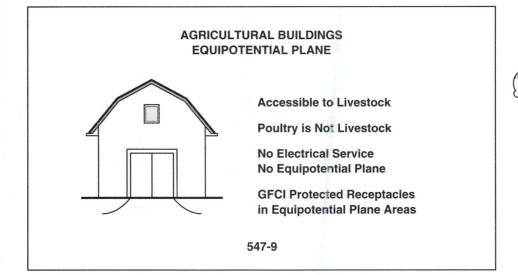

NEW RULES

This section has been rearranged, revised, and new material added. The definition for Equipotential Plane has been moved and revised.

547-9 Bonding and Equipment Plane.

(a) Definition Add *For this section livestock does not include poultry.*

Exception No. 1 An equipotential plane shall not be required where there is no electric service to the building nor metal equipment accessible to livestock that is likely to become energized.

Exception No. 2 Slated floors supported by structure that are a part of an equipotential plane shall not be required to be bonded.

(c) Receptacles. All 125-volt, 15- and 20-ampere general purpose receptacles in area having an equipotential plane shall have ground-fault circuit-interrupter protection for personnel protection.

Reason

The equipotential plane is only needed where the livestock transverses exits and entrances, and the revised definition clarifies that livestock does not include poultry. The GFCI protection is required because of the many potential grounded surfaces in the area.

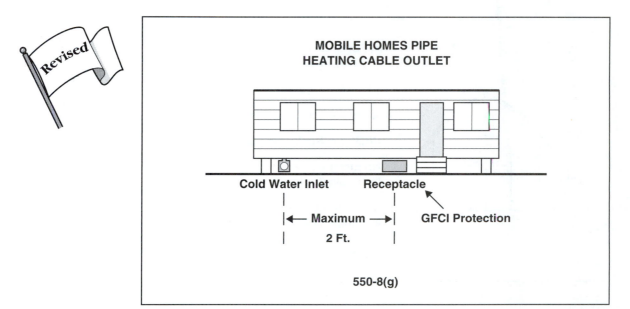

**MOBILE HOMES PIPE
HEATING CABLE OUTLET**

Cold Water Inlet   Receptacle

|←— Maximum —→| GFCI Protection

2 Ft.

550-8(g)

REVISION

The subsection title is revised and parts of the rule subdivided.

   (g) ~~*Heat Tape*~~ *Pipe Heating Cable Outlet*

(1) Location within 2 ft (610 mm) of the cold-water inlet.

(2) *Not permitted to be on the small appliance branch circuit.*

(3) *Required to have GFCI protection for personnel which is permitted to be an existing interior GFCI protection.*

(4) *Mounted on underside of mobile home* ~~*at least 3 ft from the outside edge*~~ and shall not be considered the required outdoor receptacles.

Reason

This heating tape required a safer installation. An existing GFCI protected circuit, or part of a circuit, was selected provided it is not one of the small appliance branch circuits.

Comment

This same revision is made for **Park Trailers** in *Section 552-41(d), Pipe Heating Cable Outlet.*

550-10(f) Raceways

The permitted types of raceway now include *surface metal raceway.*

MANUFACTURED HOME   550-23(b)(1)

SERVICE EQUIPMENT   *REVISED*

REVISION

Section 550-23, Service Equipment, is rearranged. The exceptions are deleted and made a part of the text. A new regulation is added: 550-23(b) *(1) The manufactured home is required to be secured to a permanent foundation.*

Reason

The revision clarifies the application of the rules and brings the *Code* into conformity with HUD requirements for a manufactured home.

Cross Check

1996	1999
	550-23 Service Equipment Title only
550-23 Mobile Home Service Equipment	**(a) Mobile Home Service Equipment**
(a) Service Equipment	Part of text
Exception No. 1	Part of text
Exception No. 2	**(b) Manufactured Homes Service**
	(1) *Permanent Foundation*
	(2) Acceptable to AHJ
	(3) Installation as per Article 230
	(4) Grounding Electrode Conductor
(b) Rating	**(c) Rating**
(FPN)	(FPN)
(c) Additional Outside Electrical Equipment	**(d) Additional Outside Electrical Equipment**
(d) Additional Receptacles	**(e) Additional Receptacles**
(e) Mounting Height	**(f) Mounting Height**
(f) ~~Grounded~~ Deleted	
(g) Marking	**(g) Marking**

ARTICLE 551

RECREATION VEHICLES AND RECREATION VEHICLE PARKS

551-30 Generator Installation

(e) Supply Conductors. This section is revised by adding regulations to cover the installation of a generator below the floor level of recreation vehicles. *The generator and its box must be installed according to its listing. The box is to be mounted at any point directly above the generator, but within 18 in. above or below the floor.*

Reason

The mounting of a generator below the floor level was not covered in the previous *Code*.

551-41(c) Ground-Fault Circuit Interrupter Protection

A new exception is added.

BASIC RULE

Where a receptacle installed to serve the countertop is within 6 ft of the sink or lavatory, it is required to be GFCI protected for personnel.

NEW EXCEPTION

Exception No. 3: De-energized receptacles that are within 6 ft (18.3 m) of any sink or lavatory due to the retraction of the expandable room section are not required to have GFCI protection.

Reason

These receptacles are not intended to serve the countertop.

551-47(a) Wiring Systems

~~Rigid metal conduit, intermediate metal conduit, electrical metallic tubing, rigid nonmetallic conduit, flexible metal conduit, Type MC cable, Type MI Cable, Type AC cable, and nonmetallic sheathed cable shall be permitted.~~ *Cables and raceways installed in accordance with Articles 330 through 352 are now permitted in accordance with their applicable articles, except as otherwise specified in this article.*

Reason

This revision permits the use of electrical nonmetallic conduit, liquidtight flexible nonmetallic conduit, and surface raceways.

551-47(g) Protected

This section is revised to include the permitted new wiring methods.

Comment

This revision permits any metallic or nonmetallic cable assembly and any metallic or nonmetallic conduit or tubing to be used as a wiring method on recreational vehicles. The phrase "except as otherwise specified in this article" pertains to outdoor installations subject to moisture or physical damage. In such cases, the installation is limited to a metal conduit or tubing or rigid nonmetallic conduit.

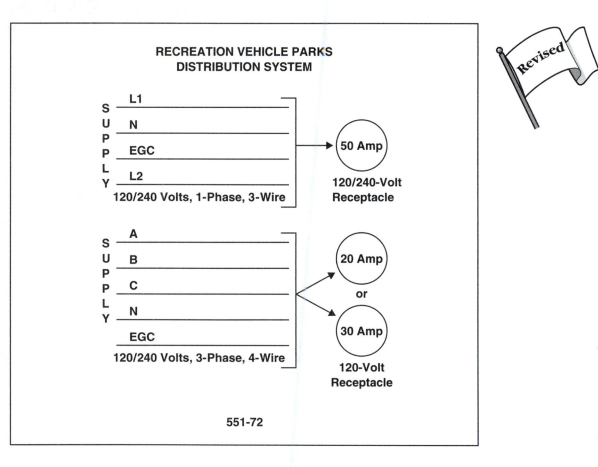

**RECREATION VEHICLE PARKS
DISTRIBUTION SYSTEM**

551-72

PREVIOUS REQUIREMENT

Recreation vehicle park receptacles were required to be supplied by a single-phase 120/240-volt, 3-wire power supply system.

REVISION

Recreational vehicle sites with 120-volt, 20- and 30-ampere receptacles are now permitted to be derived from any 120-volt, single-phase power. The neutral conductor is not permitted to be reduced in size below the size of the ungrounded conductors.

Reason

Today's recreational vehicle parks need 3-phase power that can supply single-phase power to receptacles. They also need to maintain the existing single-phase, 120/240-volt supply for 50-ampere receptacles.

ARTICLE 552

PARK TRAILERS

Article 552 was introduced into the *Code* in 1996. The revisions made in the 1999 *Code* are common to park trailers and recreational vehicles and now align the two articles more closely.

ARTICLE 555

MARINAS AND BOATYARDS

MARINAS AND BOATYARDS	555-6
FEEDERS AND SERVICES	*REVISED*

MARINAS AND BOATYARDS
FEEDERS AND SERVICES

RECEPTACLES		of the Sum of the Rating of the Receptacles		
For 1–4	100%	"	"	"
For 5–8	90%	"	"	"
For 9–14	80%	"	"	"
For 15–30	70%	"	"	"
For 31–40	60%	"	"	"
For 41–50	50%	"	"	"
For 51–70	40%	"	"	"
For 71–~~100~~ *Plus*	30%	"	"	"
~~For 101–Plus~~	~~20%~~	"	"	"

555-6

REVISION

The 20 percent line of the table is deleted and two subsections are added.

(a) Where shore power accommodations provide two receptacles specifically for an individual boat and these receptacles have different voltages (for example, one 30-ampere, 120-volt and one 50-ampere, 125/240-volt), only the receptacle with the larger kilowatt demand need be calculated.

(b) If the facility being installed includes individual kilowatt hour submeters for each slip, and is being calculated using the criteria listed in Section 555-5, the total demand amperes may be multiplied by 0.9 to achieve the final demand amperes.

Reason

The existing regulation was too conservative. Where only one receptacle is used at any one time, only the larger of the two need be included in the calculations. A sailboat and a powerboat may use the same slip, but at different times. The sailboat may use only a 30-ampere receptacle, while the powerboat needs a 50-ampere receptacle. The use of submeters tends to lower the overall demand, thereby permitting the use of the demand factor.

ELECTRIC SIGNS AND OUTLINE LIGHTING   600-6

DISCONNECTS   *REVISED*

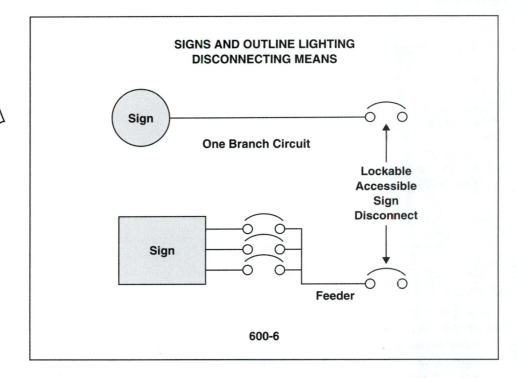

SIGNS AND OUTLINE LIGHTING
DISCONNECTING MEANS

One Branch Circuit

Lockable
Accessible
Sign
Disconnect

Feeder

600-6

REVISION

600-6 Disconnects

Each sign and outline lighting system *or feeder circuit or branch circuit supplying a sign or outline lighting* shall be controlled by an externally operable switch or circuit breaker that will open all ungrounded conductors.

Reason

This revision clarifies that when a sign is served with one branch circuit, a branch circuit disconnect is used. If the sign is served by a feeder supplying more than one branch circuit, the disconnecting means is permitted to be the feeder disconnecting means. The accessibility and the locked-in-the-open position are for safety.

SIGN AND OUTLINE LIGHTING 600-7

GROUNDING *REVISED*

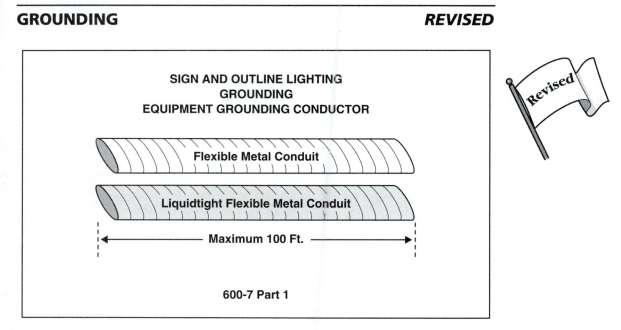

SIGN AND OUTLINE LIGHTING
GROUNDING
EQUIPMENT GROUNDING CONDUCTOR

Flexible Metal Conduit

Liquidtight Flexible Metal Conduit

◄———— Maximum 100 Ft. ————►

600-7 Part 1

REVISION

Listed liquidtight flexible metal conduit is now recognized for use as the equipment grounding conductor. Both flexible metal conduit and liquidtight flexible metal conduit when used as the equipment grounding conductor are limited to a maximum of 100 ft.

Reason

The revision is based on test data for the use of liquidtight flexible metal conduit and the 100-ft limitation.

Comment

This long paragraph details more than one requirement. So that each part is clear, it is illustrated in two parts.

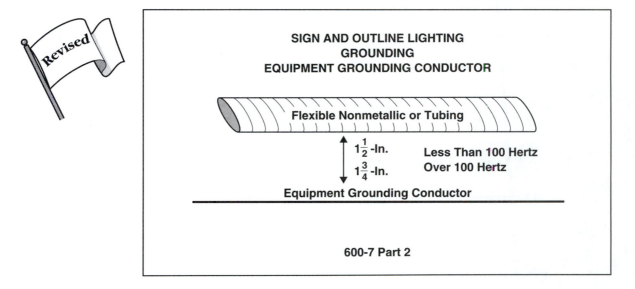

**SIGN AND OUTLINE LIGHTING
GROUNDING
EQUIPMENT GROUNDING CONDUCTOR**

Flexible Nonmetallic or Tubing

$1\frac{1}{2}$-In.   Less Than 100 Hertz
$1\frac{3}{4}$-In.   Over 100 Hertz

Equipment Grounding Conductor

600-7 Part 2

REVISION

Where flexible nonmetallic conduit or tubing is used, the equipment grounding conductor is required to be installed outside the raceway and not closer than $1^1/_2$ in. when the circuit is operating at less than 100 hertz and $1^3/_4$ in. for over 100 hertz.

Bonding conductors are required to be copper and not smaller than No. 14. Small metal parts not exceeding 2 inches in any dimension and not likely to be energized are not required to be bonded, provided they are spaced at least $^3/_4$ in. from neon tubing.

Reason

When the equipment grounding conductor was installed directly on the outside of the nonmetallic raceway, it created a situation that developed corona between the conductor and ground. This resulted in damage to the conductor in the raceway.

600-22 Ballasts

A new requirement is added to this section.

(f) Marking. A transformer or power supply must be marked to indicate that it has secondary fault protection.

Reason

The revision makes identifying proper equipment easier.

MANUFACTURED WIRING SYSTEMS

TRANSITION WIRING

604-6 Exception No. 3

NEW

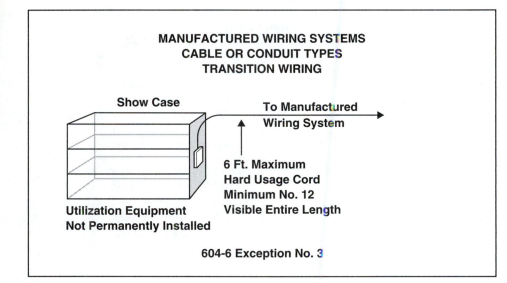

MANUFACTURED WIRING SYSTEMS
CABLE OR CONDUIT TYPES
TRANSITION WIRING

Show Case

To Manufactured Wiring System

6 Ft. Maximum Hard Usage Cord Minimum No. 12 Visible Entire Length

Utilization Equipment Not Permanently Installed

604-6 Exception No. 3

New Arrival

BASIC RULE

The basic wiring method required is armored cable or metal-clad cable, conduit, listed flexible metal conduit or listed liquidtight flexible metal conduit.

NEW EXCEPTION

Hard usage flexible cord, with No. 12 conductors not over 6 ft in length visible for its entire length, is permitted for making a transition between manufactured wiring systems and utilization equipment not permanently secured to the building or a receptacle mounted on the utilization equipment.

Reason

This exception recognizes a field practice of extending convenience power circuits from the building structure down to display cases, merchandise racks, or temporary work stations.

ARTICLE 610

CRANES AND HOISTS

A multitude of exceptions in this article are deleted and rearranged into subsection of the text. Examples are shown below.

Cross Check

1996	1999
610-11 Wiring Methods	**610-11 Wiring Methods**
Exception No. 1	**(a)** Reworded in text
Exception No. 2	**(b)** Reworded in text
Exception No. 3	**(c)** Reworded in text
Exception No. 4	**(d)** Reworded in text
Exception No. 5	**(e)** Reworded in text
610-13 Types of Conductors	**610-13 Types of Conductors**
Exception No. 1	**(a)** Reworded in text
Exception No. 2	**(b)** Reworded in text
Exception No. 3	**(c)** Reworded in text
New	*(d) Article 725 Class 1, 2, and 3 permitted*

610-41 Feeder Runway Conductors

A new subsection is added:

(a) Single Feeder

(b) More Than One Feeder Circuit. Apply the same rule that applies to single feeders.

Reason

A long runway may be supplied with more than one feeder to compensate for voltage drop.

ARTICLE 620

ELEVATORS, DUMBWAITERS, ESCALATORS, MOVING WALKS, WHEELCHAIR LIFTS, AND STAIRWAY CHAIR LIFTS

620-21 Wiring Methods

This revision adds new wiring methods.

(a) Elevator.

(2) Cars. d. Flexible metal conduit, liquidtight flexible metal conduit, liquid-tight flexible nonmetallic conduit or flexible cods and cables or conductor grouped together and taped or corded that are part of listed equipment, a driving machine or a driving machine brake are now permitted on the car assembly, in lengths not to exceed 6 ft (1.83 m) without being installed in a raceway and where located to be protected from physical damage and are of a flame-retardant type.

Reason

This revision lists the wiring methods not previously covered but permitted for equipment mounted on the elevator car assembly.

NEW PERMISSIVE RULE

Section 620-21(a)(1)a. Wiring Methods has four new exceptions **permitting** $^3/_8$ and larger liquidtight flexible nonmetalic conduit, the type with a smooth interior surface with integral reinforcement with in the conduit wall, to be used for flexible installation in lengths in excess of 6 ft for installations on:

1. Elevator cars
2. In Machine rooms and machinery spaces.
3. Escalators
4. Wheelchair lifts and stairway chair lift.

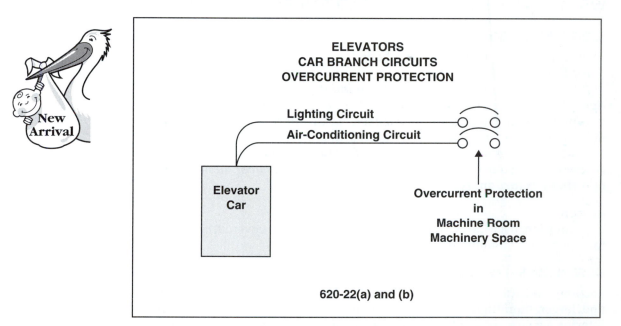

**ELEVATORS
CAR BRANCH CIRCUITS
OVERCURRENT PROTECTION**

Lighting Circuit

Air-Conditioning Circuit

Elevator Car

Overcurrent Protection
in
Machine Room
Machinery Space

620-22(a) and (b)

NEW RULE

620-22 Branch Circuits for Elevator Cars

(a) Car Light Source. *The overcurrent protection for the light source is required to be located in the machine room or the machinery space.*

(b) Air-Conditioning and Heating Sources. *The overcurrent protection for the air-conditioning and heating sources is required to be located in the machine room or the machinery space.*

Reason

This revision is made to harmonize with the Canadian Electrical Code.

ARTICLE 625

ELECTRIC VEHICLE CHARGING SYSTEM EQUIPMENT

Comment

This article first appeared in the 1996 *Code*. Since then, much has been done to develop and coordinate the use of electric vehicle charging systems and the equipment. Therefore, this article is completely revised and rearranged with numerous changes and new terminology to bring it up to date.

625-2 Definitions

The revision adds two new definitions and revises one.

Delete entire definition for "Electric Vehicle Connector" and replace it with:

Electric Vehicle Connector: *A device that by insertion into an electrical vehicle inlet, establishes an electrical connection to the electric vehicle for the purpose of charging and information exchange. This is part of the electric vehicle coupler.*

NEW – Electric Vehicle Coupler: A mating electric vehicle inlet and electric vehicle connector set.

NEW – Electrical Vehicle Inlet: The device on the electric vehicle into which the electric vehicle connector is inserted for charging information change. This is part of the electric vehicle coupler.

NEW – Personnel protection system. A system for personnel protection devices and constructional features which when used together provide protection against electric shock to personnel.

Reason

These definitions are based on *Standards for Personnel Protection Systems for Electric Vehicles (EV) Supply Circuits.*

ARTICLE 630

ELECTRIC WELDERS

Parts B and C of this article are combined into Part B with considerable rewording and revision. The original Part C was redundant except for the type of welder. The identification of the other parts are adjusted accordingly. This new Part B introduces a new method for calculating welder supply conductor sizes and overcurrent protection. Two new nameplate markings are required: 1_{1max} and 1_{1eff}. Alternatively, the welder can be marked with the rated primary current. A new equation is introduced in an FPN to illustrate 1_{1eff}.

Cross Check

1996	1999
A. General	A. General
B. AC Transformers and DC Rectifier Arc Welders	B. Arc Welders
C. Motor-Generator Arc Welders	Included in Part B. text
D. Resistance Welders	C. Resistance Welders
E. Welding Cable	D. Welding Cable

**WELDERS
SUPPLY CONDUCTORS**

| | Multiplier for Arc Welders | |
| | Nonmotor | Motor |
Duty Cycle	Generator	Generator
100	1.00	1.00
90	0.95	0.96
80	0.89	0.91
70	0.84	0.86
60	0.78	0.81
50	0.71	0.75
40	0.63	0.69
30	0.55	0.62
20 or less	0.45	0.55

630-11(a)

REVISION

Combine tables from Part B and C into one table and introduce two methods of calculations for conductors and overcurrent protection.

Method 1: ***The ampacity of the supply conductors is required to be not less than the 1_{1eff} given on the nameplate.***

Method 2: ***When 1_{1eff} is not given,*** the ampacity of the supply conductors is required to be not less than the current value determined by multiplying the rated primary current in amperes and the factor given in the table based upon the duty cycle of the welder.

Reason

The use of the 1_{1eff} value more accurately reflects the heating effect of the supply conductors because it considers both the current at idle as well as the current while welding.

Example

Method 1: 1_{1eff} given on nameplate is 38 amperes.

   Table 310-16 75°C THW copper No. 8 rated 50 amperes.

Method 2: 1_{1eff} Not given on nameplate for transformer type welder. Nameplate primary current is 65 amps, duty cycle is 70%. The 70% duty cycle from the Table Non-Motor-Generator is 0.84. Primary I × factor 65 amps × 0.84 = 54.6 amperes or No. 6 THW copper

Comment

If the I_{1eff} is not given on the nameplate, it can be calculated with the equation listed in the FPN following Section 630-12(b), provided all the values needed in the equation are known.

ARTICLE 640

AUDIO SIGNAL PROCESSING AMPLIFICATION AND REPRODUCTION EQUIPMENT

In the 1996 *Code*, this article was called "Sound-Recording and Similar Equipment." It has been completely rewritten to bring it up to date with modern technology. The article covers equipment and wiring for audio signal generation, recording, processing, amplification and reproduction systems, temporary audio and electronic organs, or other electronic musical instruments. It is subdivided into three parts. Although the article number is not changed, for practical purposes this article can be considered as completely new.

Part A. General

Part B. Permanent Audio System Installations

Part C. Portable and Temporary Audio System Installations

Section 640-2 Definitions lists seventeen definitions applicable to this article. *Section 640-3 Locations and Other Articles* is a cross reference with a multitude of other *Code* interlocking requirements.

ARTICLE 645

INFORMATION TECHNOLOGY EQUIPMENT

The title of this article was changed in the 1996 *Code* from "Electronic Computer/Data Processing Equipment" to its existing title. However, the terminology did not get changed completely in the 1996 revision, so many changes now alter "computer" or "data-processing" to "information technology."

645-2 Special Requirements for Information Technology Equipment Rooms

Delete the Fine Print Note concerning storage of materials in the information technology equipment room, as such a rule is not within the scope of the *Code*.

ARTICLE 680

SWIMMING POOLS, FOUNTAINS, AND SIMILAR LOCATIONS

680-4 DEFINITIONS

A new definition is added:

Self-Contained Therapeutic Tubs or Hydrotherapeutic Tanks. A factory-fabricated unit consisting of a therapeutic tub or hydrotherapeutic tank with all water-circulating, heating, and control equipment integral to the unit. Equipment may include pumps, air blowers, heaters, light controls, sanitizer generators, etc.

Reason

This new definition coordinates with Section 680-62, which now also covers this equipment.

SWIMMING POOLS, SPAS, HOT TUBS   680-12

EQUIPMENT DISCONNECT   *REVISED*

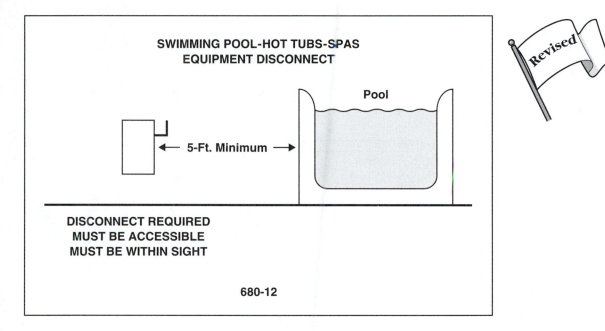

SWIMMING POOL-HOT TUBS-SPAS
EQUIPMENT DISCONNECT

Pool

← 5-Ft. Minimum →

DISCONNECT REQUIRED
MUST BE ACCESSIBLE
MUST BE WITHIN SIGHT

680-12

REVISION

680-12 Disconnecting Means A disconnecting means shall be ***provided and be*** accessible. It must be located within sight from all pool, spa, or hot tub equipment and shall be located at least 5 ft (1.52 m) ~~horizontally~~ from the inside walls of the pool, spa, or hot tub.

Reason

The revision now specifically requires a disconnecting means. The 1996 *Code* only required the disconnecting means to be accessible but did not require a disconnecting means.

680-20 Underwater Lighting Fixtures.

(b)(3) Wet-Niche Fixtures A new sentence is added. The basic rule is that fixtures are required to be bonded to the forming shell. ***Bonding is not required for fixtures listed for the application, having no noncurrent-carrying metal parts.***

Reason

This revision recognizes a product on the market that has no noncurrent-carrying metal components. There is no need or means available for grounding.

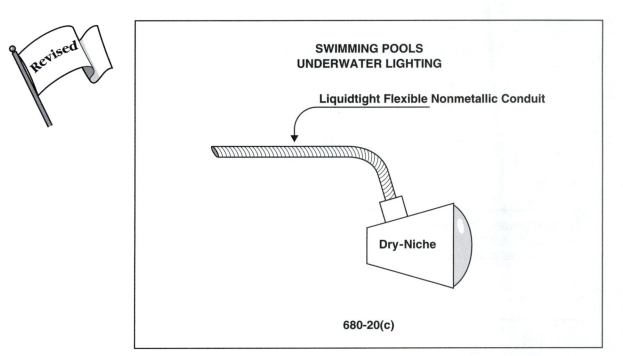

**SWIMMING POOLS
UNDERWATER LIGHTING**

Liquidtight Flexible Nonmetallic Conduit

Dry-Niche

680-20(c)

REVISION
Liquidtight flexible nonmetallic conduit is now permitted to enclose conductors supplying a dry-niche lighting fixture.

Reason
Liquidtight flexible nonmetallic conduit is listed for wet-niche fixtures in Section 680-29(b)(1) and now is recognized for use with the dry-niche fixtures.

SWIMMING POOLS   680-21(a) & (b)

JUNCTION BOXES   *NEW*

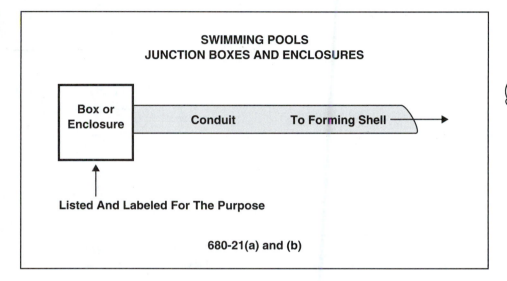

SWIMMING POOLS
JUNCTION BOXES AND ENCLOSURES

Box or Enclosure — Conduit — To Forming Shell →

Listed And Labeled For The Purpose

680-21(a) and (b)

New Arrival

NEW RULE

A junction box connected to a conduit that extends directly to a forming shell or mounting bracket or a non-niche fixture is required to be *(1) Listed and labeled for the purpose.*

Reason

Unless the box is listed and labeled for the purpose, it does not have the terminal bar for connecting the required grounding and bonding conductors.

680-21(b) Other Enclosures.

An enclosure for a transformer, ground-fault circuit-interrupter or a similar device connected to a conduit that extends directly to a forming shell or mounting bracket or a no-niche fixture is now required to be *(1) Listed and Labeled for the purpose.*

680-22 Bonding

The FPN is made a part of the text.

   It is not the intent of this section to require that the No. 8 or larger solid copper bonding conductor be extended or attached to any remote panelboard, service equipment, or any electrode, but only that it be employed to eliminate voltage gradients in the pool area as prescribed.

Reason

If a rule is the intent of the *Code*, then it should be in the enforceable part of the *Code* and not in an unenforceable FPN.

680-22(a) Bonding Parts

(1) Where reinforcing steel is effectively insulated by a listed encapsulating nonconductive compound, at the time of manufacture and installation, it is not required to be grounded.

Reason

The insulted rods are not considered to be in direct contact with the earth or concrete.

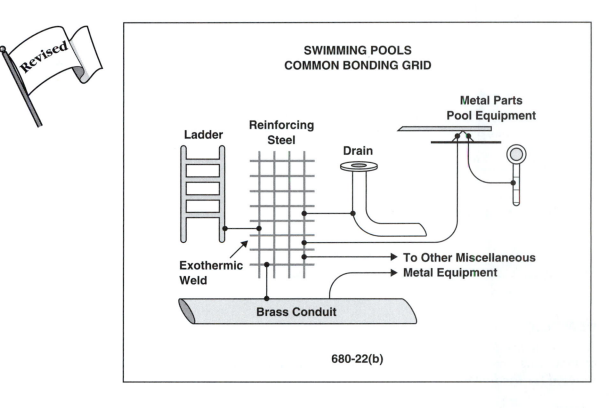

SWIMMING POOLS
COMMON BONDING GRID

Ladder · Reinforcing Steel · Drain · Metal Parts Pool Equipment · Exothermic Weld · To Other Miscellaneous Metal Equipment · Brass Conduit

680-22(b)

REVISION

680-22(b) Common Bonding Grid The parts specified in (a) above shall be connected to a common bonding grid with a solid copper conductor, insulated, covered, or bare, not smaller than No. 8. Connection shall be made by *exothermic welding or by* pressure connectors or clamps *that are labeled as being suitable for the purpose and are of the following materials:* stainless steel, brass, copper, or copper alloy. The common bonding grid shall be permitted to be any of the following:

New listing: *(4) Rigid metal conduit or intermediate metal conduit of brass or other identified corrosion resistant metal conduit.*

Reason
Exothermic welding is an acceptable and successful method of making grounding connections. The brass conduit has been used as part of the common grid and is considered safe for this purpose.

Comment
The phrase "labeled as being suitable for the purpose" indicates that there is a product on the market identified for this particular use and that it should be used.

SWIMMING POOLS   680-25(d)(2)

METHODS OF GROUNDING   *NEW*

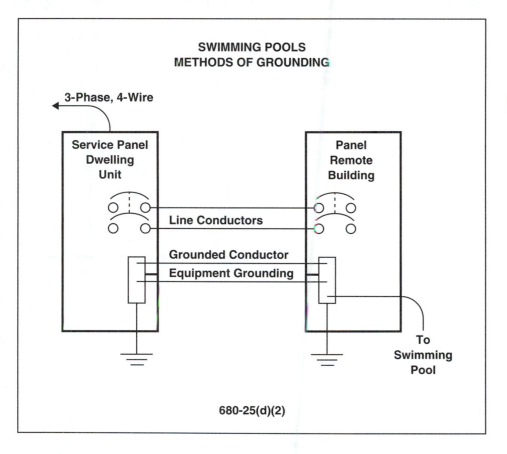

SWIMMING POOLS
METHODS OF GROUNDING

3-Phase, 4-Wire

Service Panel Dwelling Unit

Panel Remote Building

Line Conductors

Grounded Conductor

Equipment Grounding

To Swimming Pool

680-25(d)(2)

BASIC RULE

A panelboard not part of the service equipment is required to have an equipment grounding conductor installed back to the service equipment.

NEW RULE

The new rule is an alternate method of grounding for swimming pools. *A panelboard at a separate building is now permitted to feed swimming pool equipment, if the feeder meets the requirements of Section 250-32.*

Reason

This alternate method is based on a service being 3-phase, 4-wire, and the separate building and pool are a distance from the dwelling unit.

Comment

Section 250-32 covers the regulations for the installation of a feeder to a second building. This brings the installation under the same rule as any other separate building or structure. The illustration is based on Section 250-32.

OTHER THAN SINGLE-FAMILY
DWELLING UNITS EMERGENCY SWITCH   *NEW*

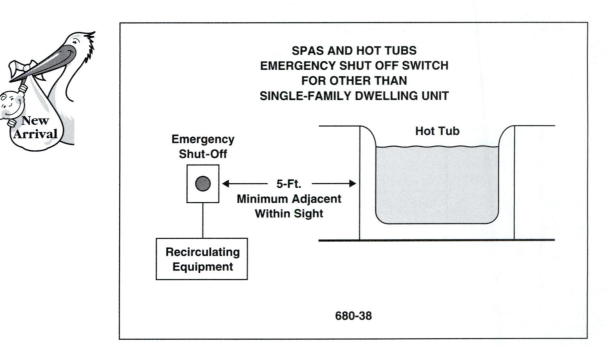

NEW RULE

680-38 Emergency Switch for Spas and Hot Tubs. A clearly labeled emergency shutoff switch for the control of the recirculation system and jet system shall be installed at least 5 ft (1.52 m) away, adjacent to and within sight of the spa or hot tub. This is not applicable to a single-family dwelling.

Reason

This regulation is a response to an accident. A person was held underwater by the discharge of a hot tub. The wording requires this switch to be installed and identified as the shutoff.

SPAS AND HOT TUBS

The first part of **Part D Spas and Hot Tubs** has one new section added, deletes the exceptions and makes them permissive rules in the text.

Cross Check

1996	1999
	680-38 Emergency Switch for Spas and Hot Tubs.
680-40 Outdoor Installations	680-40 Outdoor Installations
Exception No. 2 and 4	(a) Flexible Connections
Exceptions No. 1 and 3	(b) Bonding
680-41 Indoor Installations	680-41 Indoor Installations
(a) Receptacles	(a) Receptacle
(1)	(1)
(2)	(2)
(3)	FPN (Relocated)
FPN	(3)
(b) Lighting Fixtures, Lighting Outlets, and Ceiling Fans.	(b) Mounting Height of Lighting Fixtures, Lighting Outlets, and Ceiling-Suspended (Paddle) Fans.
(1)	(1)
Exception No. 1	Included in text of (1)
Exception No. 2	(2)

ELECTRIC SIGN   *NEW*

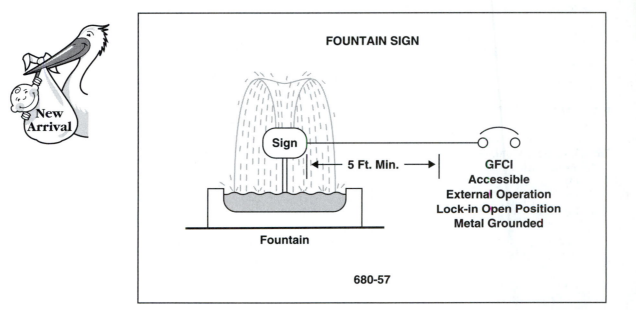

FOUNTAIN SIGN

Sign

← 5 Ft. Min. →

GFCI
Accessible
External Operation
Lock-in Open Position
Metal Grounded

Fountain

680-57

New Arrival

NEW RULE

680-57 Signs
(a) General. May be fixed or stationary to attract attention.
(b) Ground-Fault Circuit Interrupter. All circuits supplying signs are required to be GFCI protected.
(c) Location. At least 5-ft from the inside edge of the fountain.
(d) Disconnect. It must be accessible, external operation and can be locked in the open position.
(f) Bonding.
(e) Grounding. All metal parts are required to be grounded as per Article 250.

Reason

This new section coordinates with Section 600-6 for a sign located within a fountain.

NEW

680-72 *Hydromassage bathtub motors and their associated electrical connections must be accessible without damaging the building structure.*

Reason

Inspection and maintenance problems occur when the equipment and connections are not accessible.

ARTICLE 690

SOLAR PHOTOVOLTAIC SYSTEM   690-1

SCOPE   *REVISED*

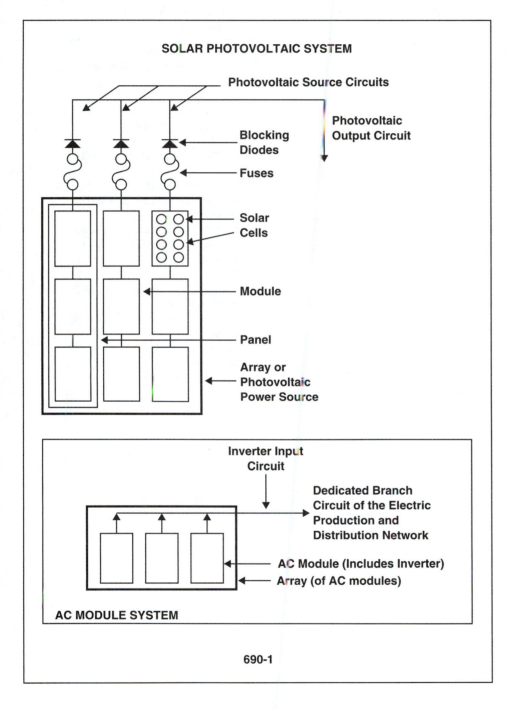

REVISION

The scope of this article is revised to update the regulations to current solar photovoltaic system technology. The entire article has been updated with new and revised material. The following is a cross check of the first two sections showing where the new material is introduced, what is revised, and illustrating the magnitude of the changes throughout the article.

Cross Check

1996	1999
690-1 Scope **Fig. 690-1**	**690-1 Scope** Revised **Fig. 690-1** Revised
690-2 Definition **Module)** **Array** **Blocking Diode** **Distribution** **Interactive System** **Inverter** **Inverter Input Circuit** **Inverter Output Circuit** **Module** **Panel** **Photovoltaic Output Circuit** **Photovoltaic Power Source** **Solar Cell** **Solar Photovoltaic System** **Stand Alone System**	**690-2 Definition** **AC Module (AC Photovoltaic** New **Array** **Blocking Diode** **Charge Controller** New **Electrical Production and** **Network** New **Hybrid System** New **Interactive System** Revised **Inverter** Revised **Inverter Input Circuit** Revised **Inverter Output Circuit** Revised **Module** Revised **Panel** **Photovoltaic Output Circuit** Revised **Photovoltaic Power Source** **Photovoltaic Source Circuit** New **Solar Cell** **Solar Photovoltaic System** **Stand Alone System** Revised **System Voltage** New

ARTICLE 695

FIRE PUMPS

Article 695, Fire Pumps, has been completely revised and rearranged to bring it up to date with other codes and modern technology.

FIRE PUMPS	695-3(a)
INDIVIDUAL SOURCES	***REVISED***

```
                    FIRE PUMPS
            ELECTRICALLY MOTOR DRIVEN

  INDIVIDUAL SOURCES

    1-Utility Service Connection  ──────▶
    2-On-Site Production Facility ──────▶      Motor      Fire
    3-Feeder Source               ──────▶                 Pump

                    695-3(a)
```

REVISION

695-3 *Power Sources(s) for Electric-Driven Fire Pumps. Electric motor-driven fire pumps shall have a reliable source of power.*
(a) Individual Source
  (1) Electric Utility Service Connection
    • *Separate service, or*
    • *Tap ahead of service but not located within the same cabinet, enclosure, or vertical switchboard section as the service disconnecting means and installed as service entrance conductors.*
  (2) On-Site Power Production Facility. The source facility is required to be located and protected to minimize the possibility of damage by fire.

Reason

This revision clarifies the requirements for the installation of an individual electrical source for electric motor-driven fire pumps.

■ ■ ■ ■ ■ ■ ■ ■ ■ ■ ■ ■ ■ ■ ■ ■ ■ ■ ■

EMERGENCY SYSTEMS   700-6(c)

TRANSFER EQUIPMENT   *NEW*

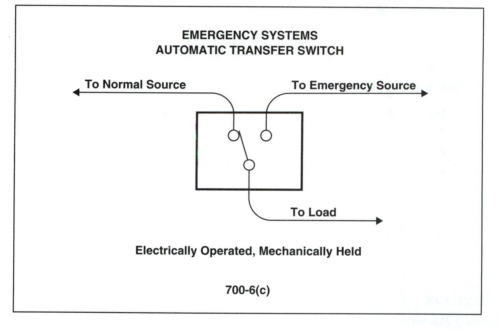

**EMERGENCY SYSTEMS
AUTOMATIC TRANSFER SWITCH**

To Normal Source   To Emergency Source

To Load

Electrically Operated, Mechanically Held

700-6(c)

NEW RULE

700-6(c) *Automatic transfer switches shall be electrically operated and mechanically held.*

Reason

An automatic transfer switch can drop out when a coil fails.

700-9 Wiring

This section is revised and rearranged, and exceptions deleted and restated as part of the text without a change in the intent of the rule.

Cross Check

1996	1999
700-9 Wiring Emergency System	**700-9 Wiring Emergency System**
(a) **Identification**	(a) **Identification**
(b) **Wiring**	(b) **Wiring** Revised
Exception No. 4	Part of text
Exception No. 1	(1)
Exception No. 2	(2)
Exception No. 3	(3)
Exception No. 5	(4)
	(c) **Wiring Design and Location** New
(c) **Fire Protection**	(d) **Fire Protection**
(1) **Feeder**	(1) **Feeder**
	a.
	b.
	c.
	d.
	e.
	f.
(2) **Equipment**	(2) **Equipment**
FPN No. 1	FPN

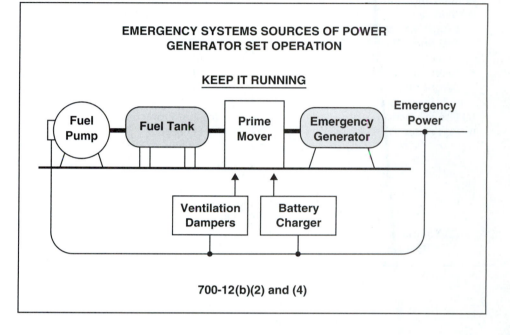

NEW RULE

(b) Generator Sets. *Where any of the following are needed to maintain the operation of the prime mover for an emergency system, they are required to be supplied by the emergency power system.*
- *Fuel pump for prime mover's day tank.*
- *Battery charger when battery's required for operation*
- *Damper used for generator ventilation.*

Reason

Every effort should be made to make sure that the emergency system does not fail because of lack of the prime mover support system.

CLASS 1 CIRCUITS   725-24 Exception No. 3

ELECTRONIC SOURCE—OVERCURRENT PROTECTION   *NEW*

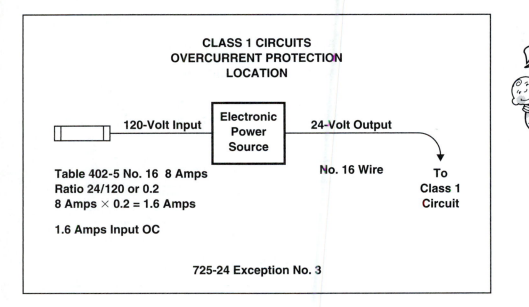

**CLASS 1 CIRCUITS
OVERCURRENT PROTECTION
LOCATION**

120-Volt Input   Electronic Power Source   24-Volt Output

Table 402-5 No. 16 8 Amps
Ratio 24/120 or 0.2
8 Amps × 0.2 = 1.6 Amps

1.6 Amps Input OC

No. 16 Wire   To Class 1 Circuit

725-24 Exception No. 3

BASIC RULE

The overcurrent protection is required to be located at the point where the conductor receives its supply.

NEW EXCEPTION

Overcurrent protection for the output conductors for a Class 1 circuit is permitted to be on the input side of the electronic power source when all the following conditions are met:
- *It is a listed electronic power source other than a transformer.*
- *The output is single-phase 2-wire.*
- *Is not considered protected by input overcurrent device.*
- *The input overcurrent device does not exceed output conductor ampacity multiplied by output-to-input voltage ratio.*

Reason

This new exception recognizes new technology not previously covered by the *Code*.

CLASS 2 AND CLASS 3 CIRCUITS 725-54(a)(1)

INSTALLATION WITH OTHER CONDUCTORS *REVISED*

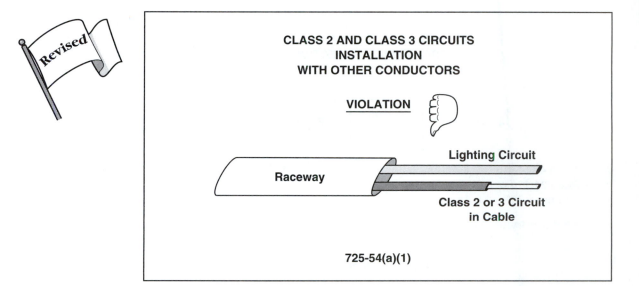

CLASS 2 AND CLASS 3 CIRCUITS INSTALLATION WITH OTHER CONDUCTORS

VIOLATION

Raceway

Lighting Circuit

Class 2 or 3 Circuit in Cable

725-54(a)(1)

REVISION

Cables and conductors of Class 2 and 3 circuits shall not be placed in any cable, cable tray, enclosure, and so forth with conductors of light, power, Class 1, and so forth.

Reason

The previous statement only mentioned conductors. The intent is that no Class 2 or Class 3 circuit conductor or cable be installed with light, power, Class 1, or similar circuit conductors. Conductors in a cable do not protect the Class 2 or Class 3 circuits from the influence of the other current-carrying conductors.

REVISION

725-54(d) Support of Conductors. An additional sentence is added to this subsection.

Class 2 and Class 3 circuit conductors shall not be permitted to be strapped, taped, or attached by any means to the exterior of any conduit or other raceway as a means of support.

Reason

The revision makes very clear that Class 2 and Class 3 circuit conductors within a building or extending beyond a building are not to be supported by a raceway.

Comment

This revision is also made for Fire Alarm circuit conductors.

CLASS 2 AND 3 CIRCUITS TABLE 725-61

CABLES USES AND PERMITTED SUBSTITUTIONS *REVISED*

CLASS 2 AND CLASS 3 CIRCUITS CABLE USES AND PERMITTED SUBSTITUTIONS

Cable Type	Use	References	Permitted Substitutions
CL3P	Class 3 Plenum Cable	725-61(a)	CMP
CL2P	Class 2 Plenum Cable	725-61(a)	CMP, CL3P
CL3R	Class 3 Riser Cable	725-61(b)	CMP, CL3P, CMR
CL2R	Class 2 Riser Cable	725-61(b)	CMP, CL3P, CL2P, CMR, CL3R
PLTC	Power-Limited Tray Cable	725-61(c) and (d)	
CL3	Class 3 Cable	725-61(b), (e), and (f)	CMP, CL3P, CMR, CL3R, CMG, CM, PLTC
CL2	Class 2 Cable	725-61(b), (e), and (f)	CMP, CL3P, CL2P, CMR, CL3R, CL2R, CMG, CM, PLTC, CL3
CL3X	Class 3 Cable, Limited Use	725-61(b) and (e)	CMP, CL3P, CMR, CL3R, CMG, CM, PLTC, CL3, CMX
CL2X	Class 2 Cable, Limited Use	725-61(b) and (e)	CMP, CL3P, CL2P, CMR, CL3R, CL2R, CMG, CM, PLTC, CL3, CL2, CMX, CL3X

TABLE 725-61

REVISION

Table 725-61 is revised by deleting several of the cable types and realigning the table.

Figure 725-61 Cable Substitution Hierarchy is revised by deleting several of the cable types and realigning the table with the following addition in the notes to the table.

Type MP-Multipurpose Cable *(coaxial cables only)*

Reason

The realigning of the table makes for easier reading. The deleted cable types are not now considered as substitution types. Other deleted types, such as FPLP, are deleted, because the only difference between FPLP and CL3P is the label on the jacket.

ARTICLE 727

INSTRUMENT TRAY CABLE TYPE ITC

This article is rearranged and renumbered, and exceptions are deleted and reworded in the text with new additions. The following illustrates the rearrangement. This article was introduced in the 1996 *Code* and now needs some fine tuning in the 1999 *Code*.

Cross Check

1996	1999
727-1 Definition **727-7 Other Articles**	**727-1 Scope** New **727-2 Definition** **727-3 Other Articles** Revised
727-2 Uses Permitted 　(1) 　(2) 　(4) 　(4) 　(5) 　(6)	**727-4 Uses Permitted** 　**(1)** In cable tray 　**(2)** In raceways 　**(3)** Hazardous locations 　**(4)** Open wiring as per 727-6 New 　　*Exception No. 1* New 　　*Exception No. 2* New 　**(5)** Aerial cable on messenger 　**(6)** Identified for direct burial 　**(7)** Under raised floors
727-3 Uses Not Permitted 　**(a) Power, Lighting, and Nonpower Limited** 　**(b) Other Circuits** 　*Exception No. 1* 　*Exception No. 2*	**727-5 Uses Not Permitted** 　Included in text 　Included in text 　*Exception No. 1* 　*Exception No. 2*
727-4 Construction 　*Exception* **727-5 Marking** **727-6 Ampacity**	**727-6 Construction** 　Included in text **727-7 Marking** **727-8 Allowable Ampacity**
727-8 Bends	**727-9 Overcurrent Protection** New **727-10 Bends**

INSTRUMENT TRAY CABLE 727-9

OVERCURRENT PROTECTION *NEW*

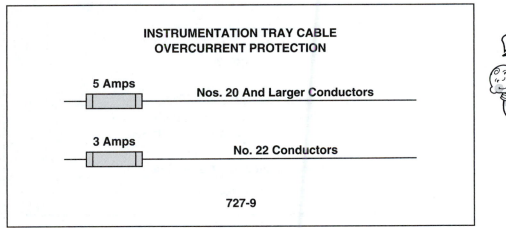

NEW RULE

Overcurrent protection shall not exceed 5 amperes for Nos. 20 and larger conductors and 3 amperes for No. 22 for conductors in Type ITC cable.

Reason

Previous requirements did not address overcurrent protection of the conductors.

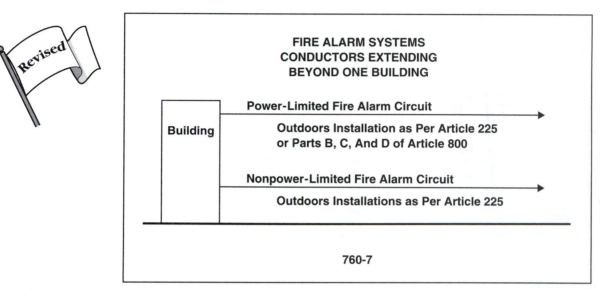

REVISION

Power-limited Fire Alarm circuits that extend beyond one building and ***run outdoors*** shall either meet the requirements of ***Parts B, C, and D of*** Article 800 ~~and be classified as communications circuits~~ or shall meet the requirements of Article 225. ***Nonpower-limited fire alarm circuits that extend beyond one building and are run outdoors shall meet the requirements of Article 225.***

Reason

This revision makes clear that even though the circuits are run outdoors as communication circuits, they are still fire alarm circuits.

Comment

Article 800 is Communication Circuits. Article 225 is Outside Branch Circuits and Feeders, and it covers conductors installed outside between two buildings.

NEW DEFINITION

A new definition for Type CI cable is given in Section 760-2. Its use is covered in Section 760-31(f), and its marking requirements are covered in Section 760-31(g.)

760-2 Definitions. *Fire Alarm Circuit Integrity (CI) Cables: Cables used in fire alarm systems to ensure continued operation of critical circuits during a specified time under fire conditions.*

760-31(f) *Fire Alarm Circuit Integrity (CI)Cable. Type CI cables are suitable for use in fire alarm systems to ensure survivability of critical circuits during a specified time under fire conditions and shall be listed as circuit integrity (CI) cable.*

760-31(g) NPLFA Cable Marking. Cables that are listed for circuit integrity are identified with the suffix "CI" as defined above. Example of marking is NPLF-CI.

760-71(i) Cable Markings. Cables identified for circuit integrity shall have a "CI" suffix marking on the cable.

Reason

This definition is based on new technology being used in the fire alarm industry. The cable is recognized by the National Fire Alarm Code.

Comment

NPLFA stands for nonpower limited fire alarm cable.

ARTICLE 770

OPTICAL FIBER CABLES AND RACEWAYS

770-2 DEFINITIONS

A new section called "Definitions" is inserted following the scope section, so the following sections are renumbered. The new terms defined are:
- *Point of Entrance*
- *Exposed*
- *FPN references Article 100 also*
- *Optical Fiber Raceway*

Section 770-6 Cable Trays

This section is moved to **770-53, Applications of Listed Optical Fiber Cables and Raceways,** and is listed under **(e) Cable Trays.**

OPTICAL FIBER CABLE AND RACEWAYS   770-6 Exception

RACEWAY SYSTEM   *NEW*

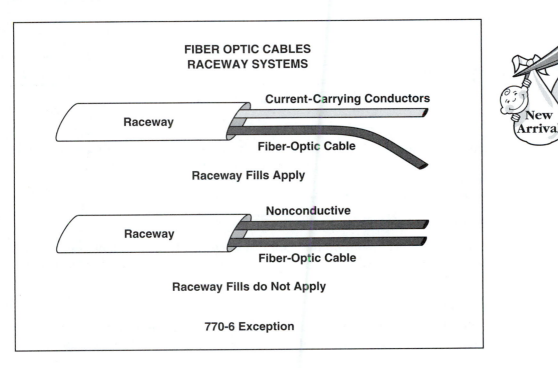

FIBER OPTIC CABLES
RACEWAY SYSTEMS

Current-Carrying Conductors

Raceway

Fiber-Optic Cable

Raceway Fills Apply

Nonconductive

Raceway

Fiber-Optic Cable

Raceway Fills do Not Apply

770-6 Exception

BASIC RULE

Raceways are permitted to be a type listed and installed in accordance with Chapter 3. Chapter 3 limits the cross-sectional area of a raceway fill, as per Chapter 9, that the conductor can fill the raceway.

NEW EXCEPTION

Unlisted underground or outside plant construction plastic innerduct is required to be terminated at its entrance to the building.

Where optic fiber cables are installed in a raceway with current carrying conductors, the fill requirements of Chapter 9 are required.

Where nonconductive optic fiber cables are installed in a raceway without current-carrying conductor, the fill requirements of Chapter 3 and 9 do not apply.

Reason

Unlisted innerduct is stopped at the entrance to the building as the raceway may not have the appropriate fire safety requirements.

When electrical current-carrying conductors are installed in the same raceway with optic fiber cables, the heat produced by the current-carrying conductors raises safety issues. When nonconductive optic fiber cables are installed in a raceway without any current-carrying conductors, there are no safety concerns because there is no heat producing current flow.

CHAPTER 8
COMMUNICATION SYSTEMS

■ ■

ARTICLE 800

COMMUNICATION CIRCUITS

REVISION

800-2 Definitions Two FPNs containing definitions followed Section 800-30. The two FPNs are deleted, rephrased, and defined in Section 800-2.
• *Block,*
• *Exposed with a FPN referring to Article 100.*

REVISION

Where describing communication conductors, the term ~~flame retardant~~ is deleted and replaced with **resistant to flame spread** because this term is more technically correct.

REVISION

Table 800-53, Cable Substitution, is revised with the addition of a column identifying **use**, and another listing the section that refers to the cable.

COMMUNICATIONS   800-48

RACEWAY SYSTEMS   *NEW*

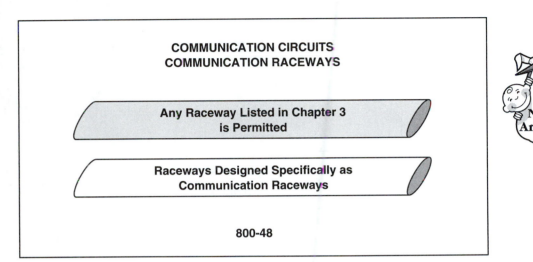

**COMMUNICATION CIRCUITS
COMMUNICATION RACEWAYS**

Any Raceway Listed in Chapter 3
is Permitted

Raceways Designed Specifically as
Communication Raceways

800-48

New
Arrival

NEW RULE

800-48 Raceways for Wiring and Cables. The raceway is permitted to be a type listed in Chapter 3 and installed in accordance with Chapter 3.

NEW EXCEPTION

Communication raceways designed specifically for communication conductors or cables but not listed in Chapter 3 are permitted but are required to be installed according to specific requirements listed in Chapter 3 for raceways.

Reason

This new rule establishes a raceway system identified for communication conductors or cables, and in so doing recognizes a nonmetallic raceway system designed specifically for communication conductor or cables. This communications raceway is not listed or recognized elsewhere in the *Code*.

Comment

Several interlocking issues explain the introduction of the new term "communications raceway systems." Section 800-48 is the basic rule. This permits the use of any raceway wiring method listed in Chapter 3. Section 800-51, Requirements for Communication Raceways, in subsections (j), (k), and (l) lists the different types of nonmetallic raceway designed specifically for communications conductors or cables. The exception to Section 800-48 permits these communication raceways and further requires them to be installed as required by Chapter 3. The specific requirements listed in Chapter 3 involve the use of proper fittings, bends, sizing, and boxes and the protection of conductors.

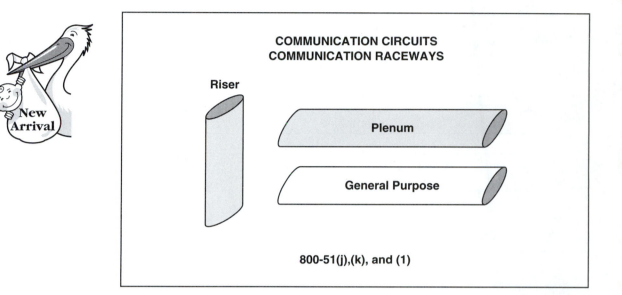

COMMUNICATION CIRCUITS
COMMUNICATION RACEWAYS

Riser

Plenum

General Purpose

800-51(j),(k), and (1)

NEW RULE

800-51 Listing Requirements for Communication Raceways.

(j) Plenum Communication Raceway. Plenum communication raceway is a nonmetallic raceway designed for communications only and listed as having adequate fire-resistant and low smoke-producing characteristics.

(k) Riser Communication Raceways. Riser communications raceway is a nonmetallic raceway designed for communications only and listed as having adequate fire-resistant characteristics capable of preventing the carrying of fire from floor to floor.

(l) General Purpose Communication Raceway. General Purpose communication raceway is a nonmetallic raceway designed only for communications and listed as having adequate fire-resistant characteristics.

Reason

These new communication raceway systems need identification.

ARTICLE 830

NETWORK-POWERED BROADBAND COMMUNICATIONS SYSTEMS

New Article

Article 830, Network-Powered Broadband Communications Systems, is a new article to cover network-powered broadband communications systems that provide any combination of voice, audio, video, data, and interactive service through a network interface unit.

A typical network-powered broadband communications system includes a cable supplying power and a broadband signal to a network interface unit that converts the broadband signal to the component signals. Typical cables are:
(1) Coaxial cable with both the broadband signal and the power on the center conductor.
(2) Composite metallic cable with a coaxial member for the broadband signal and a twisted pair for power.
(3) Composite optical fiber cable with a pair of conductors for power.

Larger systems may also include network components, such as amplifiers, that require network power.

Outdoor network-powered broadband communications systems under the exclusive control of the utility company or in building spaces used exclusively for the system are not covered by this article.

Reason

This article discusses modern technology, which the *Code* must cover.

CHAPTER 9
TABLES

■ ■

Chapter 9 Tables

The word *Examples* is deleted from the title of Chapter 9 because all examples in Chapter 9 are relocated to *Appendix D, Examples.*

Reason

Chapter 9 is an enforceable part of the *Code.* The examples were very confusing and difficult to enforce. The examples are moved to *Appendix D,* which is an informational part of the *Code.* The example numbers are re-identified to indicate they are in Appendix D, such as *D1.*

MOVE

Delete Note 10 to Table 1 and restate it as FPN No. 2 to Table 1. This note refers to the jamming of conductors when they are pulled into a raceway.

Reason

This is an informational fact rather than an enforceable *Code* rule.

Table 4 Dimensions and Percent Fill of Conduits and Tubings

This table, which consists of twelve separate tables, has an extra column added listing the conduit sizes, making the table much easier to read.

Table 5 Dimensions of Insulated Conductors and Fixture Wire

There is one value correction change in this table. The approximate area in square inches of Type TFFN wire is 0.0072.

APPENDIX D
USER-FRIENDLY

■ ■

MOVE

A new *Appendix D* is added. The examples listed in Chapter 9 are moved to the new *Appendix D.*

Reason

These examples are informational, not enforceable *Code* rules.

REVISION

Motor Circuit Conductors, Overload Protection, and Short-Circuit and Ground-Fault Protection This example has been revised by breaking it into parts a and b, rewording it, and rearranging the example for clarity.

MOVE

Mobile Homes 550-13 Calculations, ~~Example~~ This example is deleted in Section 550-13, Calculations, and is relocated as **Example D11, Mobile Home,** in Appendix D.

    Park Trailers 552-47 Calculations. The example is deleted in Section 552-478, Calculations, and is relocated as **Example D12, Park Trailers,** in Appendix D.

Reason

These examples are informational, not enforceable *Code* rules.

NEW APPENDIX E

Appendix E is added. It contains cross references for Article 250: one cross reference from the 1996 *Code* to the 1999 *Code* and one cross reference from the 1999 *Code* to the 1996 *Code.*